WONDERBOOK OF FANTASTIC FOLD-OUT FACTS

WONDERBOOK OF FANTASTIC FOLD-OUT FACTS

Includes fantastic facts about shark attacks,

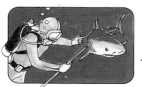  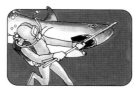

bats, lightning, bullets and guns, the Venus fly trap, killer submarines, sumo wrestlers, space exploration, deep sea diving, the bionic man, and much more

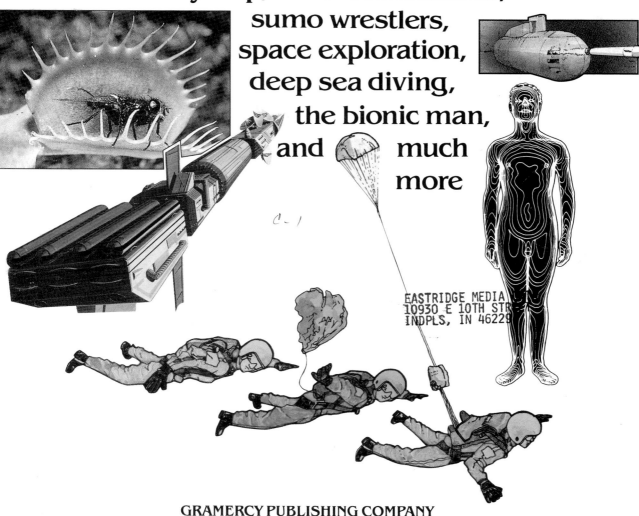

GRAMERCY PUBLISHING COMPANY
NEW YORK

This 1989 edition is published by Gramercy Publishing Company, distributed by Crown
Publishers Inc., 225 Park Avenue South, New York, New York 10003, by arrangement
with Orbis Publishing Ltd.

Printed and Bound in Hong Kong

Library of Congress Cataloging-in-publication Data

   Wonderbook of fantastic fold-out facts: an encyclopedia of amazing information
     on everything from skydiving to how to survive in the desert.
      p.   cm.
    Summary: Presents facts on nature, science, high technology, and daring
adventures on foldout pages.
    ISBN 0-517-68812-3
    1. Science—Miscellanea—Juvenile literature. [1. Science—Miscellanea.]
I. Gramercy Publishing Company.
Q163.W645   1989
  500—dc20
89-7444
CIP
AC

ISBN 0-517-68812-3
h g f e d c b a

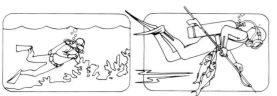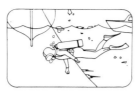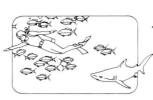

CONTENTS

WONDERBOOK OF FANTASTIC FOLD-OUT FACTS

12,000 feet, 80 miles an hour, you're hanging on the side of an aircraft. Nothing between you and the ground, way, way down; can't even see the cars. The slipstream is buffeting, pulling, pushing, taking your breath away. Two other blokes climb out alongside you, pushing you to the end of the narrow rails on which your grip on life depends. A knot of backs and heads edge out of the cabin door. One of them flexes in, then out, then in again. That's your signal. Jump! You leap into the wind. Use it. That's your control. Fly it.

The air's rushing by at 120 miles an hour and it whips your skin. Down below the other guys are forming up into a blob. You'll have to move to join up. Haul those arms back, close the legs. Must be doing 160, but not fast enough. Everything tucked in and you're into a head-first vertical dive, clocking 200 miles an hour and catching the blob real fast.

Now 'crab' that airflow; you could almost grip the stuff. Feel yourself slow back to 120. Real good. Haul back the arms and into the formation. Grip your man. You're in. Last of eight, but you're in.

SKYDIVE

A dozen Red Devil skydivers exit from the back of a Hercules at about 20,000 ft. They will free-fall for about 60 seconds, reaching up to 120 mph, flying into and out of link-up formations.

RED DEVIL'S FREE FALL KIT

Goggles
Apart from protection from the wind and cold, airborne seeds and snowflakes can be painful and dangerous at 120 mph.

Gloves
Thin gloves enable you to grip your equipment better than thick ones.

Crew knife
Necessary for cutting away tangled lines. Some skydivers carry three in case tangled lines prevent them getting hold of one.

helmet
Devils wear crash helmets, civilians can use softer and padded skull caps. Sometimes helmets are equipped with audible meters.

Cutaway pad
This houses the mechanism that releases a faulty main parachute.

Altimeter
Set before each session of jumps, this shows the diver his height above ground. In formation jumps you look at your opposite man's rather than down at your own.

Harness
Tough nylon webbing holds you and your parachutes together, giving comfortable support around the body and legs.

Parachute
Main on the bottom, reserve on the top.

Jump suit
A coverall is worn to stop any sharp edges flailing around during descent.

Warm clothes
It's going to be colder up there than it is on the ground, and when you're dropping at 120 mph you'll lose body heat fast.

Pilot chute pocket
The pilot chute is housed in here and is manually deployed. It pulls a bridle line out that releases the main chute from the backpack.

Boots
The Red Devils use supple but solid boots to give support to the ankles. However, civilian skydivers often use trainers.

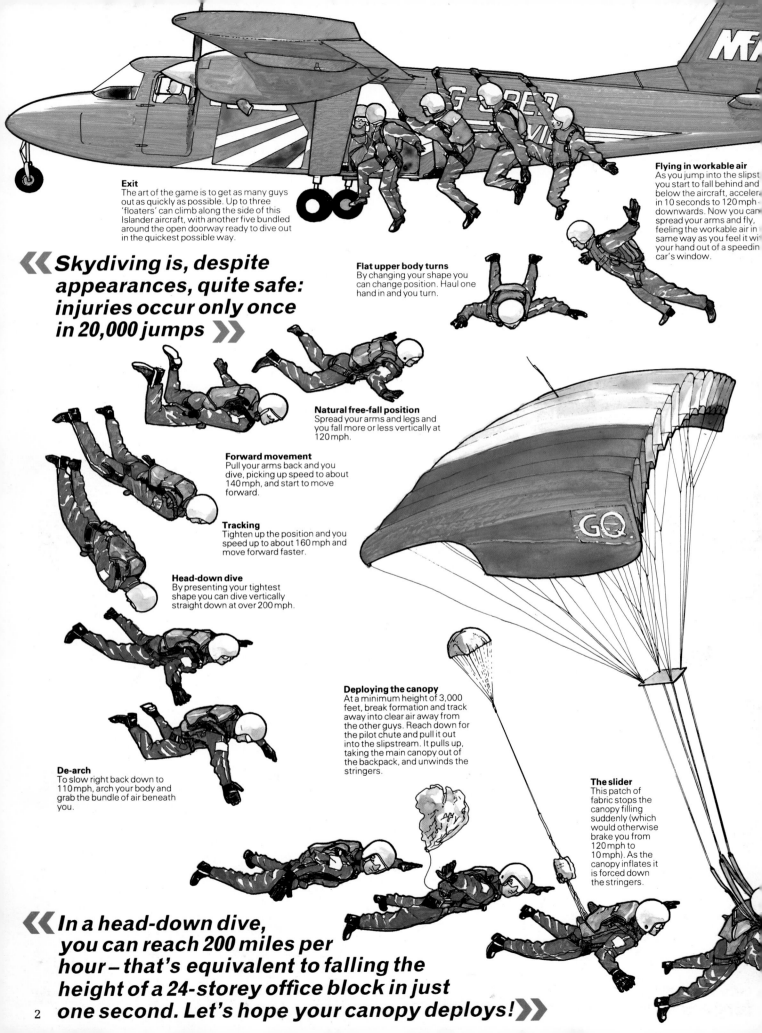

Exit
The art of the game is to get as many guys out as quickly as possible. Up to three 'floaters' can climb along the side of this Islander aircraft, with another five bundled around the open doorway ready to dive out in the quickest possible way.

Flying in workable air
As you jump into the slipst you start to fall behind and below the aircraft, acceler in 10 seconds to 120 mph downwards. Now you can spread your arms and fly, feeling the workable air in same way as you feel it wi your hand out of a speedin car's window.

« Skydiving is, despite appearances, quite safe: injuries occur only once in 20,000 jumps »

Flat upper body turns
By changing your shape you can change position. Haul one hand in and you turn.

Natural free-fall position
Spread your arms and legs and you fall more or less vertically at 120 mph.

Forward movement
Pull your arms back and you dive, picking up speed to about 140 mph, and start to move forward.

Tracking
Tighten up the position and you speed up to about 160 mph and move forward faster.

Head-down dive
By presenting your tightest shape you can dive vertically straight down at over 200 mph.

Deploying the canopy
At a minimum height of 3,000 feet, break formation and track away into clear air away from the other guys. Reach down for the pilot chute and pull it out into the slipstream. It pulls up, taking the main canopy out of the backpack, and unwinds the stringers.

The slider
This patch of fabric stops the canopy filling suddenly (which would otherwise brake you from 120 mph to 10 mph). As the canopy inflates it is forced down the stringers.

De-arch
To slow right back down to 110 mph, arch your body and grab the bundle of air beneath you.

« In a head-down dive, you can reach 200 miles per hour – that's equivalent to falling the height of a 24-storey office block in just one second. Let's hope your canopy deploys! »

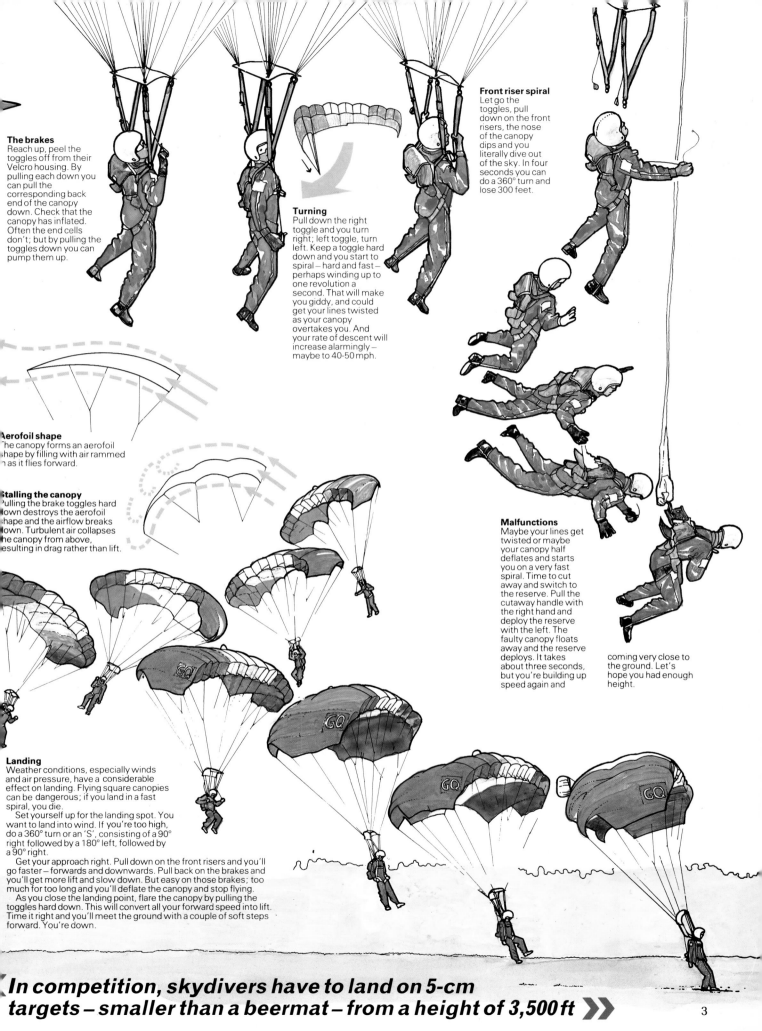

The brakes
Reach up, peel the toggles off from their Velcro housing. By pulling each down you can pull the corresponding back end of the canopy down. Check that the canopy has inflated. Often the end cells don't; but by pulling the toggles down you can pump them up.

Turning
Pull down the right toggle and you turn right; left toggle, turn left. Keep a toggle hard down and you start to spiral – hard and fast – perhaps winding up to one revolution a second. That will make you giddy, and could get your lines twisted as your canopy overtakes you. And your rate of descent will increase alarmingly – maybe to 40-50mph.

Aerofoil shape
The canopy forms an aerofoil shape by filling with air rammed in as it flies forward.

Stalling the canopy
Pulling the brake toggles hard down destroys the aerofoil shape and the airflow breaks down. Turbulent air collapses the canopy from above, resulting in drag rather than lift.

Front riser spiral
Let go the toggles, pull down on the front risers, the nose of the canopy dips and you literally dive out of the sky. In four seconds you can do a 360° turn and lose 300 feet.

Malfunctions
Maybe your lines get twisted or maybe your canopy half deflates and starts you on a very fast spiral. Time to cut away and switch to the reserve. Pull the cutaway handle with the right hand and deploy the reserve with the left. The faulty canopy floats away and the reserve deploys. It takes about three seconds, but you're building up speed again and coming very close to the ground. Let's hope you had enough height.

Landing
Weather conditions, especially winds and air pressure, have a considerable effect on landing. Flying square canopies can be dangerous; if you land in a fast spiral, you die.

Set yourself up for the landing spot. You want to land into wind. If you're too high, do a 360° turn or an 'S', consisting of a 90° right followed by a 180° left, followed by a 90° right.

Get your approach right. Pull down on the front risers and you'll go faster – forwards and downwards. Pull back on the brakes and you'll get more lift and slow down. But easy on those brakes; too much for too long and you'll deflate the canopy and stop flying.

As you close the landing point, flare the canopy by pulling the toggles hard down. This will convert all your forward speed into lift. Time it right and you'll meet the ground with a couple of soft steps forward. You're down.

In competition, skydivers have to land on 5-cm targets – smaller than a beermat – from a height of 3,500 ft ➤➤

Shark Attack

The Great White Shark moved lazily through the warm, dim waters of the Pacific like a large shadow: a tonne of streamlined muscle cruising in a large circle in search of prey, probing the gloom with an elaborate array of sensors tuned to perfection – the ultimate killing machine.

Suddenly its speed increased a little and there was an almost imperceptible change of course. The delicate pressure sensors around its cold, empty eyes and in a line along its body had picked up the faintest of vibrations in the water. Its huge head swung from side to side as it peered into the dimness that surrounded it, sampling the water passing into its nostrils for the slightest scent. The curve of its course increased slightly as the shark responded to the vibrations by sweeping in a long, slow spiral towards their source. Then it saw its target. Jaws gaping, it accelerated to nearly 25 mph before it struck. The jaws clamped shut and two fearsome arcs of razor-sharp, triangular teeth scythed together.

Robert Pamperin and Gerald Lehrer bobbed in the warm Californian waters about 200 metres from shore. It was Sunday, and the two young men were spending the afternoon diving for shellfish in La Jolla Cove. Suddenly Pamperin cried and surged half out of the water before disappearing in a welter of foam.

His companion dived down immediately, and witnessed the terrible spectacle of his friend emerging from a cloud of dark blood with only his upper body protruding from the mouth of the Great White.

On that summer's day, another name was added to the list of shark victims.

At 1500 metres from the target, the shark begins to detect vibrations and low-frequency sounds with pressure-sensitive organs in the skin and small, stone-like structures resting on sensory cells inside the head (otoliths). When the shark is alerted to the movements and noise, it turns towards the source to investigate.

《 *A shark caught in the Adriatic had a raincoat, three overcoats and a car number plate in its stomach* 》

A Great White shark displays its savage teeth, haunting eyes and snub nose. It's enough to make the most experienced of divers panic – but don't. If he hasn't decided on you for a meal he'll drift by, but any sudden movement from you could make him change his mind, and then . . .

Few sights are more likely to strike terror into the heart of a casual swimmer than the sinister dark triangle of a shark's fin slicing through the water towards him.

At first the innocent victim may think he's got away with his life, for the first contact with the attacking shark will be not the nightmare of its multi-toothed jaws but a sickening thump as the shark crashes into its intended prey and scrapes it with its flank. However, the shark has simply made its last check that it has found food, using the toothlike denticles and sensors resembling taste buds that cover its skin. Next, it will swing round in a tight turn before accelerating at up to 20 mph for the kill. In a final corkscrew motion, the shark will clamp its jaws around its victim and rip it in two.

Deadly detectors

Sharks have developed an astonishing array of sensors and detectors that guide them to food; there is scarcely an inch of their streamlined bodies that does not contribute directly to their ability to find and devour their prey. These range from radar-like vibration detectors and a highly developed sense of smell to equipment for sensing pressure changes in the water. When the shark picks up an indication that prey is about, it will often swim *around* its victim, spiralling gradually towards it for the kill.

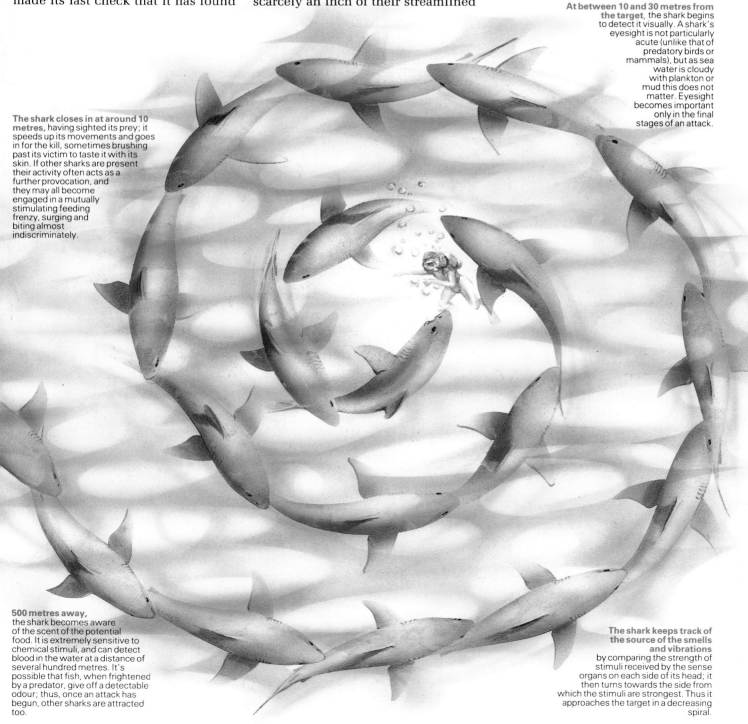

At between 10 and 30 metres from the target, the shark begins to detect it visually. A shark's eyesight is not particularly acute (unlike that of predatory birds or mammals), but as sea water is cloudy with plankton or mud this does not matter. Eyesight becomes important only in the final stages of an attack.

The shark closes in at around 10 metres, having sighted its prey; it speeds up its movements and goes in for the kill, sometimes brushing past its victim to taste it with its skin. If other sharks are present their activity often acts as a further provocation, and they may all become engaged in a mutually stimulating feeding frenzy, surging and biting almost indiscriminately.

500 metres away, the shark becomes aware of the scent of the potential food. It is extremely sensitive to chemical stimuli, and can detect blood in the water at a distance of several hundred metres. It's possible that fish, when frightened by a predator, give off a detectable odour; thus, once an attack has begun, other sharks are attracted too.

The shark keeps track of the source of the smells and vibrations by comparing the strength of stimuli received by the sense organs on each side of its head; it then turns towards the side from which the stimuli are strongest. Thus it approaches the target in a decreasing spiral.

It's possible that once a shark has targeted its prey, it locks on to it like a missile – this may explain why sharks attacking swimmers in shallow water will, six times out of seven, ignore people who come to the victim's rescue. In some cases the shark has actually brushed aside would-be rescuers in order to strike again and again at its chosen target.

The human factor

Exactly why sharks should attack people remains a mystery. Simple hunger may entice the shark into the attack in the first place, but colour or the behaviour pattern of the target may also be triggering factors. Splashing is said to drive off sharks, but on one occasion in 1936 a White shark shot past a man swimming sidestroke to go for the left leg of a 16-year-old kicking vigorously while doing the crawl.

Kind ones and killers

Not all sharks will home in with such relish on a human target; some species eat nothing but plankton. And despite the fearsome reputation of the Great White – hero of *Jaws* – it is probably the Blue Whaler that is the most dangerous.

Those who venture into the deep can only be thankful that the prehistoric shark *carcharadon megalodon* has vanished from the oceans. A fossilised tooth from this monster in the Oceanographic Museum in Monaco is 10 times the size of those of a Great White shark – which suggests that the megalodon's jaws were capable of swallowing a Ford Granada in one gulp!

SIX KILLERS

There are over 300 varieties of shark, ranging from the tiny Dwarf Shark (23 cm long) to the Whale Shark (up to 18 metres). Few pose a real danger to man; the two largest species, the Whale Shark and the 12-metre Basking Shark, eat only plankton and other tiny organisms. Others, like the Port Jackson Shark, have rounded teeth adapted for eating crabs and molluscs.

Despite the fearsome reputation of the Great White, it is actually the Blue Whaler Shark who has gained the greatest reputation as a maneater and lacks the natural caution of its larger brother. Other extremely dangerous species include the Tiger and Mako sharks as well as the strange and sinister-looking Hammerhead.

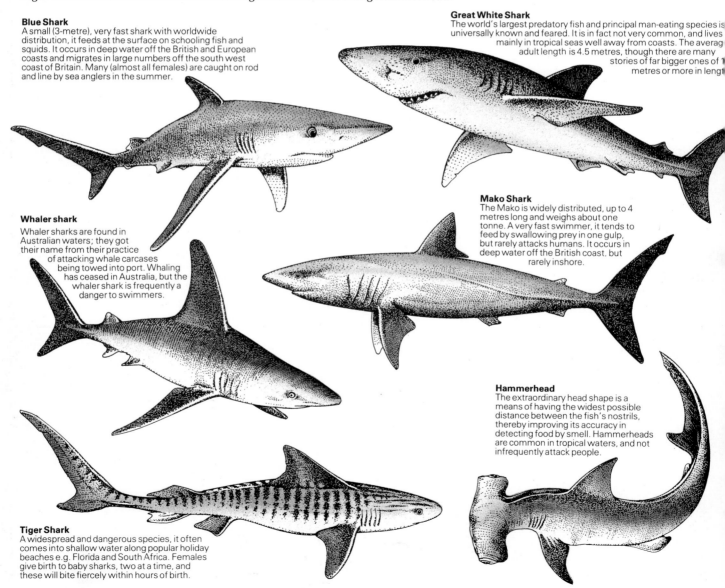

Blue Shark
A small (3-metre), very fast shark with worldwide distribution, it feeds at the surface on schooling fish and squids. It occurs in deep water off the British and European coasts and migrates in large numbers off the south west coast of Britain. Many (almost all females) are caught on rod and line by sea anglers in the summer.

Great White Shark
The world's largest predatory fish and principal man-eating species is universally known and feared. It is in fact not very common, and lives mainly in tropical seas well away from coasts. The average adult length is 4.5 metres, though there are many stories of far bigger ones of 1 metres or more in length.

Whaler shark
Whaler sharks are found in Australian waters; they got their name from their practice of attacking whale carcases being towed into port. Whaling has ceased in Australia, but the whaler shark is frequently a danger to swimmers.

Mako Shark
The Mako is widely distributed, up to 4 metres long and weighs about one tonne. A very fast swimmer, it tends to feed by swallowing prey in one gulp, but rarely attacks humans. It occurs in deep water off the British coast, but rarely inshore.

Hammerhead
The extraordinary head shape is a means of having the widest possible distance between the fish's nostrils, thereby improving its accuracy in detecting food by smell. Hammerheads are common in tropical waters, and not infrequently attack people.

Tiger Shark
A widespread and dangerous species, it often comes into shallow water along popular holiday beaches e.g. Florida and South Africa. Females give birth to baby sharks, two at a time, and these will bite fiercely within hours of birth.

« *The largest fish ever caught with rod and line was a 1208-kg white shark over 5 metres long* **»**

10 MOMENTARILY BLINDED BY THE MISSILE, THE PILOT LOOKS ROUND TO SEE THE MIL MI-24 'HIND' GUNSHIP HELICOPTER BEHIND HIM, CLOSING FAST AND READYING FOR ANOTHER MISSILE SHOT. THE PILOT PULLS THE APACHE INTO A GUT-WRENCHING TURN, THE ROTORS SLAPPING THE AIR AND CUTTING THE TREETOPS.

9 AS THE APACHE PULLS AWAY FROM THE ATTACK, THE COCKPIT IS LIT UP BY FLARE FROM A SOVIET SA-7 ANTI-AIRCRAFT MISSILE STREAKING PAST. THE MISSILE HAS NOT HOMED ON THE HELICOPTER THANKS TO ITS SPECIAL SHROUDED EXHAUST, WHICH DISSIPATES THE HEAT GIVEN OUT SO THE HEAT-SEEKING MISSILE HAS NOTHING TO AIM FOR.

11 THE TURN BRINGS 'WHISPER THREE' OUT BEHIND THE 'HIND', AND THE PILOT LETS FLY WITH A STINGER MISSILE. THE 'HIND' DOES NOT HAVE THE HEAT SHROUDING OF THE APACHE, AND THE STINGER HAS A DEADLY ACCURATE HEAT-SEEKING HEAD. WITHIN SECONDS IT SLAMS INTO THE 'HIND,' TURNING THE APACHE'S OPPONENT INTO A FIREBALL.

12 AFTER ITS EVENTFUL MISSION, 'WHISPER THREE' HEADS FOR ITS CLEARING IN THE FOREST. IT PASSES OVER A DOWNED APACHE AMONG THE TREES. DESPITE ITS AMAZING STRENGTH, THIS MACHINE HAS TAKEN TOO MUCH PUNISHMENT.

ht spots from laser beam

The laser beam has to be
eld on to the target throughout
e Hellfire's flight

13 THE STRUCTURE IS BUILT TO BE CRASHWORTHY, WITH WHEELS, ROTORS AND TAIL DESIGNED TO FOLD UP UNDER STRESS, LEAVING THE CREW COMPARTMENT INTACT.

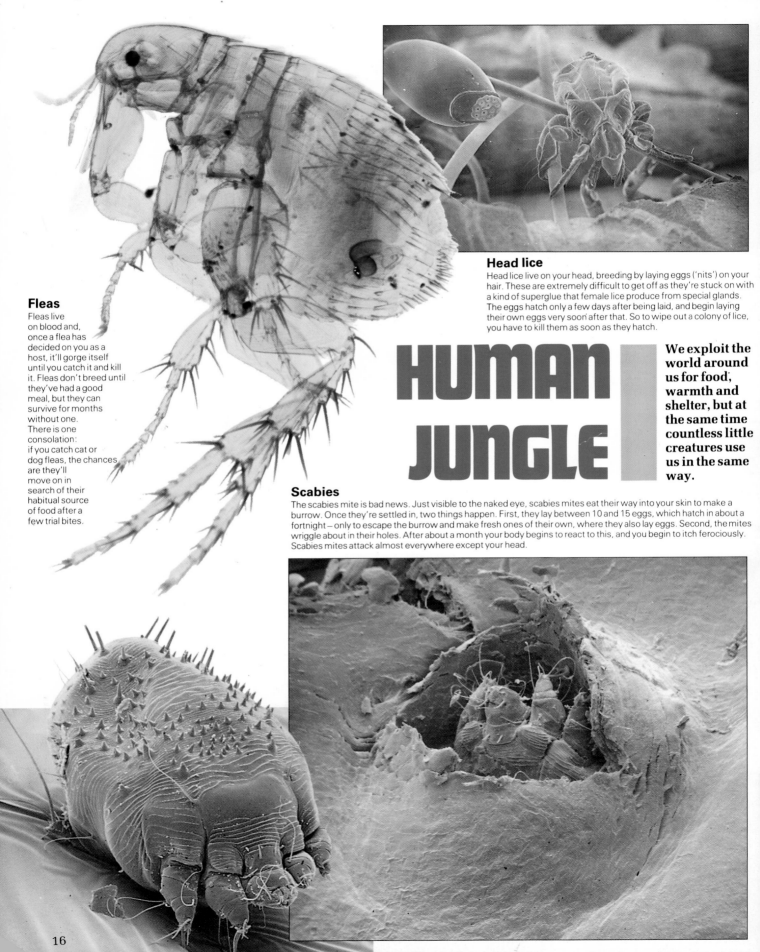

Fleas

Fleas live on blood and, once a flea has decided on you as a host, it'll gorge itself until you catch it and kill it. Fleas don't breed until they've had a good meal, but they can survive for months without one. There is one consolation: if you catch cat or dog fleas, the chances are they'll move on in search of their habitual source of food after a few trial bites.

Head lice

Head lice live on your head, breeding by laying eggs ('nits') on your hair. These are extremely difficult to get off as they're stuck on with a kind of superglue that female lice produce from special glands. The eggs hatch only a few days after being laid, and begin laying their own eggs very soon after that. So to wipe out a colony of lice, you have to kill them as soon as they hatch.

HUMAN JUNGLE

We exploit the world around us for food, warmth and shelter, but at the same time countless little creatures use us in the same way.

Scabies

The scabies mite is bad news. Just visible to the naked eye, scabies mites eat their way into your skin to make a burrow. Once they're settled in, two things happen. First, they lay between 10 and 15 eggs, which hatch in about a fortnight – only to escape the burrow and make fresh ones of their own, where they also lay eggs. Second, the mites wriggle about in their holes. After about a month your body begins to react to this, and you begin to itch ferociously. Scabies mites attack almost everywhere except your head.

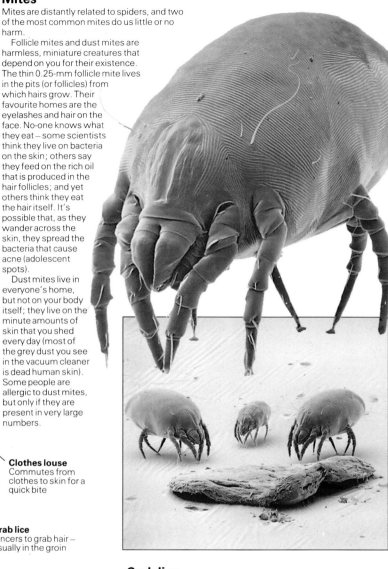

Clothes lice

Clothes lice nest in clothes but feed from the chest, shoulders and upper arms. All lice survive by piercing the skin with their sharp mouth parts and sucking blood.

Mites

Mites are distantly related to spiders, and two of the most common mites do us little or no harm.

Follicle mites and dust mites are harmless, miniature creatures that depend on you for their existence. The thin 0.25-mm follicle mite lives in the pits (or follicles) from which hairs grow. Their favourite homes are the eyelashes and hair on the face. No-one knows what they eat – some scientists think they live on bacteria on the skin; others say they feed on the rich oil that is produced in the hair follicles; and yet others think they eat the hair itself. It's possible that, as they wander across the skin, they spread the bacteria that cause acne (adolescent spots).

Dust mites live in everyone's home, but not on your body itself; they live on the minute amounts of skin that you shed every day (most of the grey dust you see in the vacuum cleaner is dead human skin). Some people are allergic to dust mites, but only if they are present in very large numbers.

SCRATCHING A LIVING

Head louse
A frequent visitor, leaving eggshells as calling cards

Follicle mite
Comfort and food in your eyelashes

Scabies mite
Three people in a thousand are the unlucky hosts

Cat and animal fleas
Guests of pets, but will try you for a meal

Clothes louse
Commutes from clothes to skin for a quick bite

Crab lice
Pincers to grab hair – usually in the groin

Dust mites
Why let dead skin flakes go to waste when they could feed a hungry mite?

« *Your eyelashes are home to over 30 mites* »

Crab lice

Crab lice infest the body, especially the groin, although they will thrive in hair in the armpits and even in your eyebrows. Because they can't fly, or even hop in the way fleas can, lice have to walk from one person to another.

OUTPACING THE COMPETITION

Dragsters can accelerate from 0-200 mph ten times faster than either Concorde or a top-of-the-range Porsche 959. By the time a dragster is bursting through the quarter-mile marker at 250 mph, a shuttle space booster has barely left the ground.

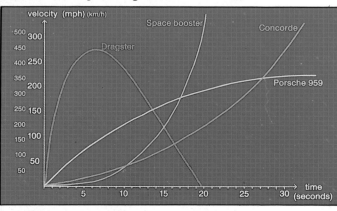

velocity (mph) (km/h)

Space booster

Dragster

Concorde

Porsche 959

time (seconds)

IN THE HOT SEAT

The tension rises as soon as you're being strapped into the cockpit, its special aroma of oil, sweat and leather in your nostrils. Then there's a gentle thump as a car with padded bumpers shoves you forward and turns over the engine.

First, up to the bleach puddle, slip the clutch, spin the wheels. The bleach boils away in a great hissing cloud as those fat back tyres soften for maximum grip.

Roll up to the start line now. Keep an eye on the temperature. You need to keep nursing the engine heat, timing it to reach peak operation at about 75°C just as the lights go green.

On the start line, hit the gas pedal as hard as you can for a second for a dry burnout. This hots up the tyres and puts some slip into the clutch. Without that slip at the start you slam power into the wheels so suddenly that they'll just spin uselessly.

Ready now. The first two yellow lights are lit up on the Christmas tree. You're in line with the track, ready for countdown – two

《 *The fastest dragsters do not run on petrol, but on a literally explosive mixture of alcohol and nitro-methane* 》

6 SECS

Flames burst from the spinning rear wheels as a dragster heats up its tyres to achieve maximum grip. Its supercharged V-8 engine will develop tremendous power – up to 3,000 horsepower – and that has to be converted through the tyres into acceleration.

THE HEART OF A DRAGSTER

Nearly all dragsters are powered by supercharged engines. By forcing larger amounts of fuel/air mixture into the engine, the supercharger develops a huge power increase over conventional engines.

Injector scoop
Air enters the engine through a 'bug-catcher' air scoop mounted above the engine block

Fuel
In Top Fuels, this is a mixture of nitromethane and methanol. It is injected into the airflow in a fine spray, forming an explosive mixture with the air

Supercharger
The fuel/air mixture is now 'supercharged', the mixture being pressurised by a belt-driven compressor powered from the crankshaft

Combustion
The super-rich, super-explosive mixture is fed from the supercharger into the hemispherical combustion chamber, where it explodes with much greater force than in normal engines, so much so that for about six seconds a typical V-8 engine will pump out up to 3,000 hp (a normal Chrysler or Chevy block puts out around 200)

Engine
The engine goes through the normal four-stroke cycle, exhausting the burnt gases to the manifold. The piston drives the camshaft and draws more fuel into the combustion chamber for the next phase of ignition

re yellow lights to flash on and off, and
n go. The trick is to 'spear the green' – hit
throttle as soon as that last yellow dims
t. Go too soon, the red light trips, and
u're out of the race.
Time it just right and you don't even see the
en as you blast away, the engine scream-
, all 3,000 horsepower surging into the
eels. Zero to 100 mph in just over a second.
orce five times that of gravity crams you
ck into the seat. Drivers have been known
black out at this point.
Another couple of seconds and pressure
s by about half. Now you're hitting about
0 mph but still getting faster. Two seconds
er you flash past the quarter-mile light at
nething over 250 mph.
Foot off the throttle now, you reach up to
go the air brake. Only then you're flung
ward in the seat as the big chute cracks
en. At last you're coasting off the strip to
et the truck with your crew. Another race
er.

ORDEAL IN THE ANDES

What would *you* do? Stuck high amid the mountains without food or adequate clothing, weak with hunger and little or no prospect of rescue – how would you survive? The survivors of a crash in the Andes had this problem, which they overcame in a startling and, to many, a disturbing fashion.

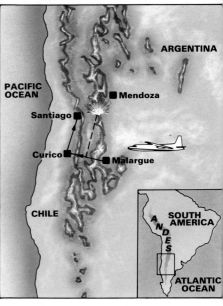

The pilot error that turned the plane north could not have been worse. Instead of safely heading for Santiago over the coastal plain of Chile, the Fairchild headed deep into the second highest mountain range on earth.

On Friday, 13 October 1972, a Fairchild FH-227 took off from Mendoza, Argentina, carrying 40 rugby players, their friends and families, and five crew. Ahead of them was a short flight, across some of the highest peaks in the Andes, to Santiago, Chile.

Halfway across the mountains the aircraft flew into dense cloud and severe turbulence. Lost amid the jagged peaks, the plane crashed into a valley 3,500 metres above sea level. Amazingly, 32 people survived. They had heard the pilots in radio contact with Chile, and reckoned they would be rescued in a day or two.

Nevertheless they carefully rationed what food they had: a dozen chocolate bars, as many packets of sweets and nuts, three bottles of wine and three of spirits. Worse, their luggage had been lost in the crash. Many of the survivors thus had to face night-time temperatures of 30° below freezing in nothing more than shirt-sleeves.

The joy of the survivors upon being rescued knew no bounds. Here, after 72 days on the mountain, a young man leaps through the high meadow to which he and his fellows have been brought by helicopter.

Planes were seen overhead, but the white-painted aircraft wasn't spotted in the snow. After 10 days the survivors heard on a salvaged transistor radio that the search had been called off. By now, their food was gone, several survivors had died, and those remaining were very weak. Gradually they realised that, if they were ever to escape, they would have to do something unthinkable. They would have to eat the bodies of their friends.

Human food

Strips of flesh were cut from the bodies and dried. Oil was collected to use as a laxative, for constipation was chronic. The only fuel was a few boxes that had held soft drinks, so the meat was eaten raw or dried except on two 'cooking days' a week. The more squeamish found the meat more palatable when fried: it tasted rather like beef.

Survival brought out unexpected qualities in people. 'Nando' Parrado had been shy and awkward, always in the shadow of his friend Panchito Abal. The crash had killed Panchito – and Nando's mother and sister. Despite this tragedy, he became a tower of strength. Calm, strong and unselfish, he hardly ever lost his temper, and weaker members of the party came to depend on him entirely.

On the night of 30 October, an avalanche killed eight more. Those left decided to send out parties to find help, but most were defeated by weakness and the cold. Parrado, Roberto Canessa and Antonio Vizintin set out again on 11 December, but Vizintin had to turn back.

Ten days later Parrado and Canessa reached a peasant's hut, and over the next two days the remaining 14 survivors were ferried to safety – and a conventional diet – in Chilean Air Force helicopters. The last to be lifted out had been in the mountains for a total of 72 days.

HOW TO SURVIVE A MOUNTAIN AIR CRASH

Mountains have sudden and dramatic changes of climate. Bright sun can cause snow blindness and burning in one moment, and cloud and wind bring frostbite and hypothermia the next. Night-time temperatures can drop to 40° below freezing, or worse.

To survive in a hostile, cold mountain environment and have a decent chance of being rescued, remember the following set of priorities.

1 Do not stay in the crashed aircraft; there is always a risk of fire, and if this doesn't happen it will act as a refrigerator. But do not stray far from the wreck, as it will be far more easily spotted than you will.

2 Attend to first aid for yourself and for your companions. Don't let anything distract you from this.

3 Next, set up a signal – or several signals – to guide rescuers. Prepare flares or a fire, or use debris, rocks or vegetation to write an easily visible SOS or a simple cross; straight lines are rare in natural landscape and are easily seen. Some kind of flag may also help, and a mirror or sheet of glass can be used for flashing heliograph signals.

4 Find or build a shelter. You may be able to use parts from your aircraft, or find a cave. If not, use branches or brushwood to build a lean-to. The base of a fir tree makes a ready-made shelter. Snow and ice holes are very snug. Insulate your shelter as thoroughly as you can, and keep it small.

5 Gather some bedding; make sure it's firm, dry and flat. Small branches from fir trees or ferns are very effective. Try and insulate yourself from cold ground by raising your bed slightly. If you have a fire, heat up some stones and take them to bed.

6 Make sure you have a water supply; look for rivers or streams, and make sure any water you have is protected from evaporation or contamination. If no water is available, collect condensation or dew. You can melt snow or ice for drinking water, but do not eat it unmelted – it will lower your body temperature and leave you even more prone to dehydration and exposure.

7 Don't drink alcohol; it increases the circulation and therefore lowers your body temperature.

8 Dress as warmly as possible; several thin layers are better than one thick one. Clothes should be windproof and waterproof, but do not wear tight plastic; your sweat will condense, soak your clothes and probably freeze. Keep your unworn clothing dry – take it to bed with you to keep it out of the cold and damp.

9 Avoid frostbite by keeping your hands, feet, ears and face well covered and out of the wind. First signs of frostbite are numbness and whiteness and by the time it starts to hurt you're in trouble. If anyone is beginning to suffer from frostbite rub his or her extremities with snow to restore circulation – rubbing without snow will break the skin tissues and cause infection.

10 Decide how long you may have to wait for rescue and ration your food and drink accordingly. Food is a low priority so be stingy – you can survive for at least two weeks without it before your energy is seriously impaired.

The worst fire disaster in a single building was an outbreak in a Chinese theatre in 1845 in which 1,670 people perished »

Great balls of fire leap up around a fire chief's helicopter in the downtown area of an American city. These air bursts are from clouds of inflammable gases that have risen from the building below and ignited in the heat of the fire.

FIRE

In the heat from the sun, a seed germinates in the ground, pushes up green shoots to trap energy from the light, and grows into a sapling. At last, a tree stands tall – a massive bundle of energy, a great store of calories from the warmth of the sun.

The tree is felled, the wood shaped, seasoned, cut to make the timber frame of a house. And then, one day, an accidental flame catches the timber. Those slumbering calories are released at last, in a violent, terrifying blaze.

Learning to make fire was one of man's greatest discoveries, as fundamental to civilisation as the invention of the wheel. Fire, if it is harnessed, can be man's greatest friend. The controlled release of energy can be put to thousands of uses: from baking a brick to burning our rubbish, from cooking our food to stripping paint.

The Greek myths say that man stole fire from the gods. Nothing could demonstrate more clearly how fire can be as fickle and destructive as any ancient deity than the kind of conflagration that destroys entire cities.

Middle Ages disasters

Medieval cities were often razed by fire, their closely packed wooden buildings and open hearths inviting devastation. London suffered twice: in 1212, when 3,000 died trapped in the houses that were on London Bridge, and in the Great Fire of 1666, which destroyed three quarters of the city and made 200,000 homeless.

The day Chicago died

In 1871 Chicago was a thriving city of 334,000 inhabitants living and working in some 60,000 buildings. On the night of 7 October, a major fire broke out in the city's timber-yards and spread rapidly, thanks to high winds and the many wooden buildings that existed among newer, stone structures. The entire fire-fighting force of the city fought the blaze throughout the night and for most of the following day. Many were injured and, by the time the fire was put out, everyone was exhausted.

A few hours later, an old woman was tending her sick cow by the light of a kerosene lamp. The cow kicked the lamp over – and in seconds the

A major fire in Chicago in 1871 exhausted the city's firefighting forces – and when a second, unrelated outbreak followed only hours later, people fled in panic and the city centre was left in ruins.

barn was ablaze. By the time weary firemen reached the scene, more than 50 buildings were burning. Vast clouds of sparks and embers carried the fire from house to house. The heat was so intense that whole structures were exploding instantly into flames.

The fire spread through the city so fast that the fire crews were overtaken by the flames and had to abandon their machinery. People stampeded in utter terror, trampling others underfoot as huge billows of flame blasted through the streets consuming everything in their path. Within 24 hours, nearly 10 sq km of the city centre were devastated; 250 people had perished, 80,000 were homeless, and 17,000 buildings had been destroyed. The cost of the damage was estimated at $200 million.

High-rise horror

Today, such large-scale conflagrations would seem to be behind us. Cities are constructed of fire-resistant materials, modern fire-fighting is a sophisticated science, and radio communications bring a rapid response to any such emergency. Nevertheless, the terrible prospect of a major urban fire is always with us.

On Christmas Day, 1971, a fire broke out in Seoul, South Korea, in a 22-storey hotel that was only two years old. It started on the second floor, when a gas burner exploded in the coffee shop, and spread rapidly upwards through the stairwell – which, ironically, had been designed to help occupants escape in just such an emergency.

The fire service reacted swiftly but found that their ladders could reach

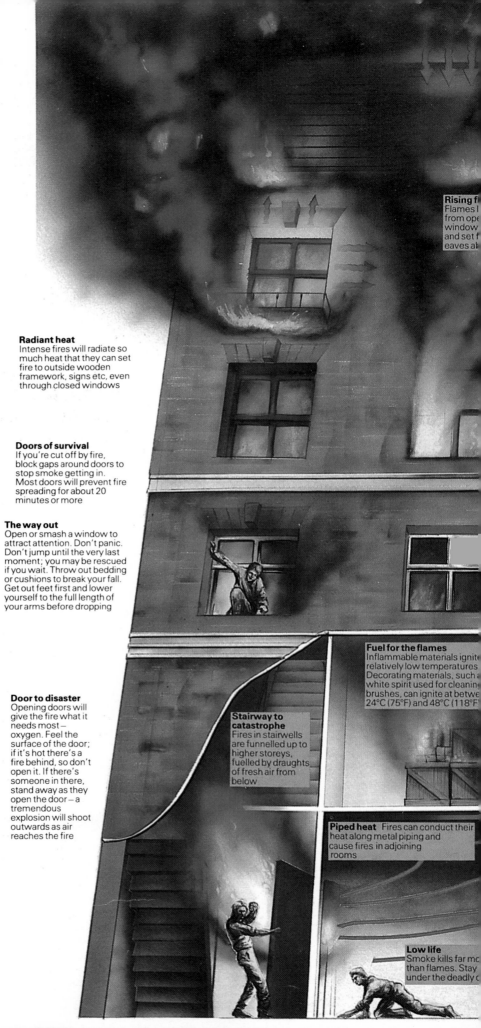

Rising f[...]
Flames l[...]
from ope[...]
window [...]
and set f[...]
eaves ab[...]

Radiant heat
Intense fires will radiate so much heat that they can set fire to outside wooden framework, signs etc, even through closed windows

Doors of survival
If you're cut off by fire, block gaps around doors to stop smoke getting in. Most doors will prevent fire spreading for about 20 minutes or more

The way out
Open or smash a window to attract attention. Don't panic. Don't jump until the very last moment; you may be rescued if you wait. Throw out bedding or cushions to break your fall. Get out feet first and lower yourself to the full length of your arms before dropping

Door to disaster
Opening doors will give the fire what it needs most – oxygen. Feel the surface of the door; if it's hot there's a fire behind, so don't open it. If there's someone in there, stand away as they open the door – a tremendous explosion will shoot outwards as air reaches the fire

Stairway to catastrophe
Fires in stairwells are funnelled up to higher storeys, fuelled by draughts of fresh air from below

Fuel for the flames
Inflammable materials ignite[...] relatively low temperatures[...] Decorating materials, such a[...] white spirit used for cleanin[...] brushes, can ignite at betwe[...] 24°C (75°F) and 48°C (118°F[...]

Piped heat Fires can conduct their[...] heat along metal piping and cause fires in adjoining rooms

Low life
Smoke kills far mo[...] than flames. Stay [...] under the deadly [...]

A small, careless accident, such as the dropping of a lighted cigarette, can lead to catastrophe. This is the terrible chain of events that can take place once a fire has started, often resulting in death.

spread of fire
ce the roof space is alight fire will
el along the timbers. Fire drifting
r parapets can radiate heat
nwards and set fire to adjoining
dings

Collapse
Fire in a roof space or upper storeys will cause the ceilings to collapse, and these can take lower floors with them

gs of safety
red ceilings are a good insulation against the spread of fire, but timber or ceilings can ignite and transfer the fire to rooms above

ath
ill build up
ually filling
from the
wnwards.
minutes
ignition the
kely to be
, and after 3
he dense,
s smoke will
where

Heat high
Intense heat builds up in the higher parts of the room, reaching 1500°F (815°C) within a couple of minutes. At a temperature of 900°F (482°C) or more, other surfaces and furniture will ignite on contact. Aerosols will explode, their highly inflammable contents showering across the room like napalm

y gas
enough air, the furniture
rst into flames, filling the
with thick, black smoke,
gases and particles of
g plastic

1500°
1200°
900°
600°
180°

Channels of fire
Hot fumes can work their way into cavities and up into roof spaces

Seats of fire
Modern furniture is often made of highly inflammable petroleum-based materials. A most common cause of fire is cigarette ends dropped into the folds of furniture where they smoulder, giving off poisonous fumes

only to the eighth floor. Hundreds of guests, screaming for help from the windows, were beyond help. In desperation, many people clung to mattresses and jumped – only to die on the street below. Of the 300 guests and staff, fewer than half were saved.

Brazilian office blaze

In Sao Paulo, Brazil, on 1 February 1974, fire broke out in a faulty air conditioning unit on the 12th floor of a new 25-storey office block in which 600 people were working. The relatively fireproof structure of the building did not prevent hundreds of square metres of carpet, partitioning, furnishings and ceiling tiles from bursting into flames, gutting the entire floor.

« In February 1983 bush fires swept across the Southern Australian outback faster than an express train, and in 48 hours killed 70 people and hundreds of thousands of animals »

As in the Korean disaster, the emergency stairwells funnelled the blaze up to the next floor, and then the next, until the upper storeys were all ablaze. Yet again, rescue equipment was unable to reach the burning floors, and by the time the inferno was brought under control 200 people were dead and another 200 were injured. Although many had managed to reach the roof, helicopters were able to give only limited help, as they were buffeted by the fierce updraughts caused by the intense heat, and pilots could scarcely see through the smoke. Sixty people perished within sight of salvation.

Harbour lights

The most violent source of fire is an explosion. In 1944, the merchant ship *Fort Stikene* put in to Bombay harbour, carrying 1400 tonnes of high explosive and a cargo of cotton. A fire started in the hold, and confused attempts were made to stop the burning. But, as the temperature in the holds rose, an explosion ripped the stern from the ship, killing 40 firefighters and setting fire to neighbouring vessels. Another ship crashed into the *Fort Stikene* and a second explosion blew both ships into fiery fragments, sending burning bales of cotton raining down on to warehouses and other ships. One vessel of 4000 tonnes was lifted out of

'IT SPREAD SO FAST...'

On 11 May 1985, Bradford City were playing their final match of the season against Lincoln City at their home ground, Valley Parade. Just before the end of the first half, someone dropped a lighted match or cigarette in block G of the old wooden main stand. Falling through the floorboards, it set fire to over 70 years of debris accumulated below – programmes, paper cups and the like – that within 30 seconds ignited adjacent timbers. Within a minute, the fire was large enough for one of the policemen on duty to ask for assistance, and two minutes later smoke forced the spectators towards the front and rear of the stand. In less than five minutes, a panic-stricken crowd was spilling out on to the pitch. Tragically, 56 people were to lose their lives – most being trapped in the hellish inferno of the access corridor at the back of the stand.

In five minutes 43 seconds, smoke drifting under the lip of the centre of the stand showed that the underside of the roof was well alight, with the horror of molten tar added to the intense heat

In six minutes 3 seconds, the north end of the stand was totally ablaze, with choking smoke billowing out over the pitch. In the heart of the inferno was the narrow corridor to the exits that ran along the rear of the stand

In six minutes 35 seconds, the inferno was consuming the entire wooden structure of the main stand. Normally, it took up to 10 minutes to clear the stand after a match – and this picture was taken less than seven minutes after the fire's insignificant start.

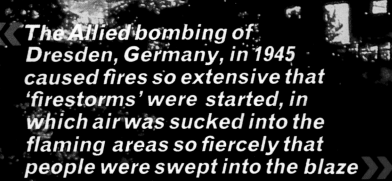

The Allied bombing of Dresden, Germany, in 1945 caused fires so extensive that 'firestorms' were started, in which air was sucked into the flaming areas so fiercely that people were swept into the blaze

the water by an explosion, blown over a 15 metre-high building and deposited in flames on to the quayside. An area of 40 hectares was gutted, and some fires smouldered for four months. Nine hundred died and 2,000 were injured.

Leaping flames

Forest and brush fires are natural phenomena, but these are often sparked by an unthinking human. Cigarette ends and discarded bottles (which magnify the sun's rays) are frequent causes. Once these fires get going, they can be driven by winds at awesome speeds, destroying everything in their path as they sweep across the countryside. The worst forest fires occurred in Wisconsin and Michigan in the USA in 1871, on the same day as the fire that destroyed Chicago. The fire started in several places, and was pushed by gale-force winds that turned the sky orange with burning fragments. Twenty-three towns were burned out in its path, 1,500 people losing their lives.

Other fires can be equally terrifying, especially those in crowded and enclosed spaces like discos, rail carriages, football stadiums or cinemas. In these blazes, poisonous fumes or mass hysteria can often be more deadly than the fire itself.

A tragic example of this occurred in Paisley, Scotland, in 1929. The local cinema was packed with children when a fire started in the projection room. The cinema management reacted quickly and calmly, and threw the burning film into the street, where it lay harmlessly smouldering. But enough smoke got into the auditorium for someone to shout 'Fire!'

minutes 41 seconds, the roof was entire...
rough and the whole wooden structure was...
The intensity of the heat had charred the ground...
of the stand, and anyone venturing close at this...
would have found the metal railings blazing...

THE DANGEROUS TRIANGLE

An outbreak of fire is like the meeting of three sides of a deadly triangle.

One side of the triangle is OXYGEN, found in the air and in the chemical composition of some solid materials.

The second is FUEL, which may be a solid substance such as wood, paper or plastic, a liquid such as petrol or alcohol, or a gas such as hydrogen.

The third is HEAT. This converts the fuel to the gaseous state that allows chemicals to combine and, hence, to combust.

Remove any one of these ingredients, and the fire will go out.

OXYGEN — FUEL — HEAT

other warehouse goes up in smoke – just one of ...illion building fires that plague America each ...r. In a similar incident the bituminous roof of a ...neral Motors warehouse caught fire, caved in ...d destroyed the entire 14-hectare building – the ...e of 24 football pitches.

In the uproar that followed, 70 children were trampled to death.

Man's fickle friend

In theory, our knowledge of fires and their prevention is advanced enough to ensure that large-scale disasters can never happen. In practice, commercial interests, financial constraints and, most often, simple human error virtually guarantee that almost any building, modern or otherwise, is a potential host to fire – mankind's everyday friend and occasional, catastrophic, foe.

FIRE! FIRE!

Close doors and windows to limit the spread of flames, smoke and fumes

*

Do not pause to collect belongings or valuables

*

Stay as low as possible, and keep well below the smoke layer

*

Use a moistened cloth to cover your nose and mouth and protect your lungs – many people die from the effects of smoke rather than from the fire itself

*

Use available fire appliances as quickly as possible but allow ample time for escape

*

If in the immediate vicinity of the flames, remove any highly inflammable clothing, particularly those made from synthetic materials

*

ABOVE ALL – DO NOT PANIC! More lives are lost for this reason than almost any fire would otherwise account for

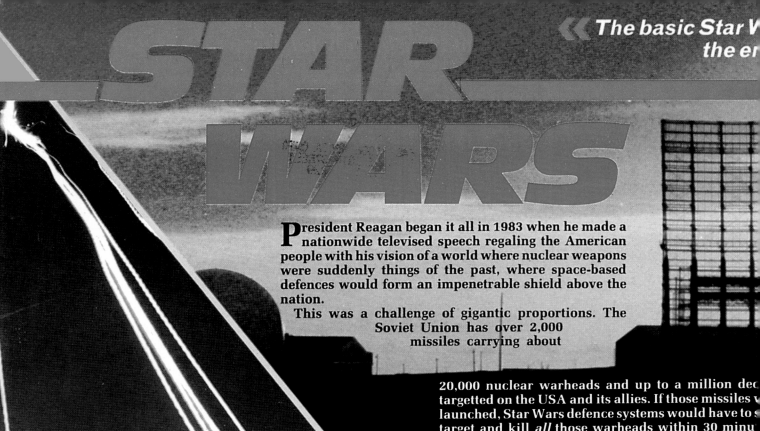

« The basic Star W
the er

President Reagan began it all in 1983 when he made a nationwide televised speech regaling the American people with his vision of a world where nuclear weapons were suddenly things of the past, where space-based defences would form an impenetrable shield above the nation.

This was a challenge of gigantic proportions. The Soviet Union has over 2,000 missiles carrying about 20,000 nuclear warheads and up to a million dec targetted on the USA and its allies. If those missiles v launched, Star Wars defence systems would have to s target and kill *all* those warheads within 30 minu while they were travelling at 10,000 miles an hour ac an area of some 50 million square kilometres, throug billion cubic kilometres of space. And that's just to tect America!

Machine management

The battle would have to be fought by compute humans are far too slow. Their programs would be credibly complicated – about 10 million lines long, would have to be *absolutely* error-free. The last minutes of a Space Shuttle launch have a prog of 'only' 90,000 lines, and think how often run into delay and difficulty.

The system would include weapons are, as yet, only dreams: not even so tifically-valid theories, still less pr machines. There would be huge ser 36,000 kilometres up to detect l ches, and strings of super-sens satellites to track enemy miss and communicate back to battle-management compu

The threat–multiple warheads streak through the atmosphere at the end of their path. They are travelling at about 10,000 mph.

« Star Wars systems will have to patrol 20 billion cubic kilometres of space looking for targets the size of dustbins **»**

A Titan 34 launche pair of surveillance satellites from Space Launch Complex 6 at

test of a high-energy chemica

...he huge **AN/FPS 50** early-warning
...dars (more than 50 metres high) will be
...upplemented by space-based systems
...hat will detect missiles at the moment of
...aunch.

...per beams

...nt space-based laser and particle beam weapons
...uld direct the energy sufficient for the needs of a small
... into a pinpoint beam, and aim it at targets the size of
...tbins thousands of kilometres away travelling at
...n per second.

...mb power

...etic energy weapons would fire 60 plastic projectiles
... second at speeds of 100 km per second. X-ray lasers,
...vered by exploding hydrogen bombs, would in the
...ant of their own destruction fire a huge pulse of
...rgy through 50 crystal rods over a distance of
...000 km. Land-based lasers each consuming the out-
...of a large nuclear power station would project light
...ms onto giant orbiting mirrors and onwards to their
...gets.

...Vill it work? Nobody knows. There can be no trials. It
... have to work first time – or else . . .

...side a **US** Air Force
...ntrol centre, the
...erations board is
...pdated while
...chnicians man the
...dar screens.

A FLAG missile (Flexible Agile
Guided) is launched. These
intercept targets in-atmosphere,
so cannot use nukes.

One of the first steps in any
space war will be to blind the
enemy's satellite 'eyes'. The
Americans have developed an
anti-satellite missile (ASAT),
launched from the F-15 fighter.

America's first ABM
(Anti-Ballistic
Missile) system was
developed in the
1960s using nuclear-
armed Spartan
missiles to blow up
warheads in the
upper atmosphere.

In spite of tragedy, the **US** Shuttle programme
is too important to the Strategic Defense
Initiative to be allowed to stop. Even so, the
grounding of Shuttles after the Challenger
disaster has been crippling.

The one thing that
'Star Wars'
systems may
have difficulty in
handling are
submarine-
launched missiles
fired at short range
on a low traject...

The nose cone of this
ICBM is carrying
three warheads
(which are about the
size of dustbins).
Imagine hitting those
from more than
1,000 km away!

Decoys and re-entry vehicles
Many of the decoys are nothing more than aluminium balloons, which reflect radar just like genuine RVs. In addition to the 10 or more warheads carried by larger missiles, there may be as many as 50 decoys. If the missiles are not taken out in the boost phase, then the 2,000 targets in the early stages of a massive first strike will multiply up to something like 100,000 or even millions of targets during the mid-course phase.

RV bus
The re-entry vehicles (RVs) are carried into orbit on a platform in the missile nose cone. This 'bus' also carries large numbers of decoys and penetration aids (penaids) designed to fool anti-missile defences.

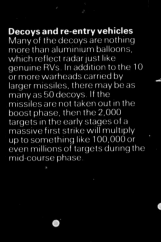

STAR WARS
THREE PHASE DEFENCE

American Star Wars defences would attack incoming Soviet missiles and warheads in each of three phases: the **Boost Phase** *as enemy missiles are rising from their launch platforms, the* **Post-boost Phase** *when the missiles are releasing their cargoes of warheads and decoys on the way to their targets, and the* **Terminal Phase** *when surviving warheads re-enter Earth atmosphere heading for the targets.*

...launched kinetic ...terceptor
...ar ground-launched ...have to be accurate ...make direct hits on ...han 2 metres long, ...at speeds of up to ...ph. The HOE, or ...verlay Experiment, ...from Kwajalein in the ...cific, successfully made ...gle interception in June ...nst an incoming ...The 'sunray' that ...o allow more chance of ...as unnecessary — the ...t the target dead centre.

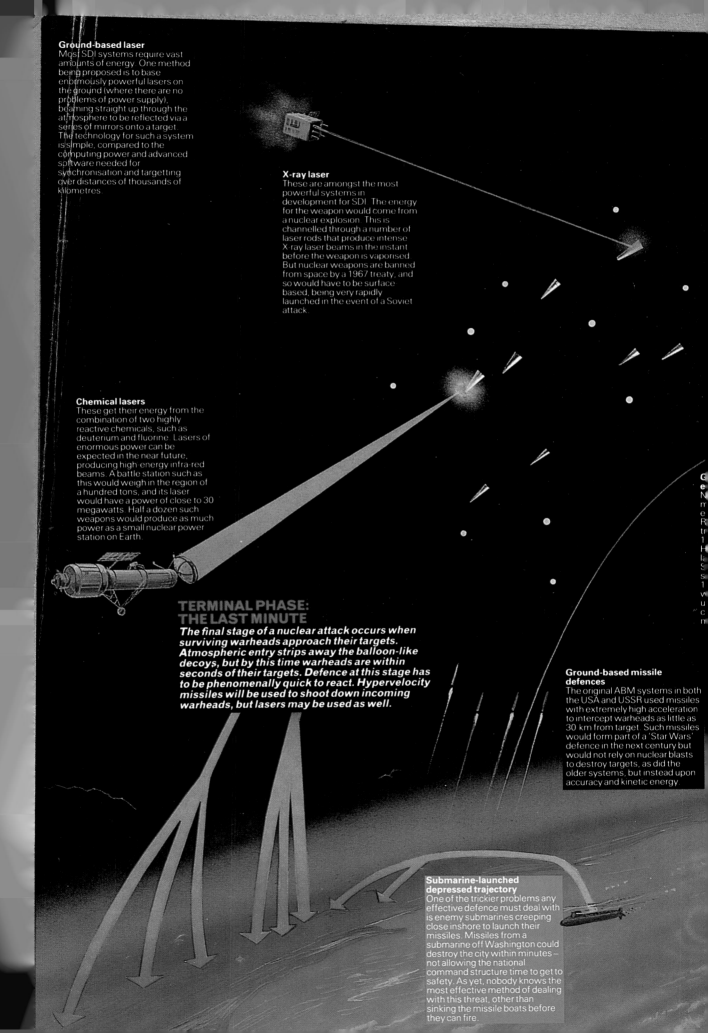

Ground-based laser
Most SDI systems require vast amounts of energy. One method being proposed is to base enormously powerful lasers on the ground (where there are no problems of power supply), beaming straight up through the atmosphere to be reflected via a series of mirrors onto a target. The technology for such a system is simple, compared to the computing power and advanced software needed for synchronisation and targetting over distances of thousands of kilometres.

X-ray laser
These are amongst the most powerful systems in development for SDI. The energy for the weapon would come from a nuclear explosion. This is channelled through a number of laser rods that produce intense X-ray laser beams in the instant before the weapon is vaporised. But nuclear weapons are banned from space by a 1967 treaty, and so would have to be surface based, being very rapidly launched in the event of a Soviet attack.

Chemical lasers
These get their energy from the combination of two highly reactive chemicals, such as deuterium and fluorine. Lasers of enormous power can be expected in the near future, producing high-energy infra-red beams. A battle station such as this would weigh in the region of a hundred tons, and its laser would have a power of close to 30 megawatts. Half a dozen such weapons would produce as much power as a small nuclear power station on Earth.

TERMINAL PHASE: THE LAST MINUTE

The final stage of a nuclear attack occurs when surviving warheads approach their targets. Atmospheric entry strips away the balloon-like decoys, but by this time warheads are within seconds of their targets. Defence at this stage has to be phenomenally quick to react. Hypervelocity missiles will be used to shoot down incoming warheads, but lasers may be used as well.

Ground-based missile defences
The original ABM systems in both the USA and USSR used missiles with extremely high acceleration to intercept warheads as little as 30 km from target. Such missiles would form part of a 'Star Wars' defence in the next century but would not rely on nuclear blasts to destroy targets, as did the older systems, but instead upon accuracy and kinetic energy.

Submarine-launched depressed trajectory
One of the trickier problems any effective defence must deal with is enemy submarines creeping close inshore to launch their missiles. Missiles from a submarine off Washington could destroy the city within minutes — not allowing the national command structure time to get to safety. As yet, nobody knows the most effective method of dealing with this threat, other than sinking the missile boats before they can fire.

STAR WARRIOR COUNTDOWN

Some time in the not too distant future . . .

'Get me the Key Holes of the Siberian missile bases. We've got what looks like a mobile missile deployment along the Trans Siberian railroad.' Lt Gen Murray of USAF Space Command had been looking at the images generated by the HALO infra-red satellite array 36,000 km above the Indian Ocean. Switching to 'Key Hole', a system of high-resolution satellites, would give far more detail on the sites themselves.

He was worried. First the Soviets had pushed their ballistic missile subs close to the US coast. Now these missile deployments made things look really bad. Time to inform the President.

The vital codes were received in reply. Systems were go. The vast SDI circus was going into operation. Satellites to monitor thousand-missile launches, laser battle stations to wage war miles up in space, space mirrors to reflect laser light thousands of miles, rail guns, hypervelocity missiles. Nobody knew how they'd work for real. But they would all have to. There was no time for mere human decision-making. Events would run so fast that the computers would have to have full control.

Laser Battle Station

As the space systems were powered up, Murray looked at the general situation display screen on the wall. He noted that the Space Launch Complex at Vandenberg AFB, California, was in the process of launching a dozen of the H-bomb powered X-ray laser battle stations. His face whitened. If they were putting *those* babies up, then this was going to be big. His eyes strayed to the Naval Operations display, monitoring Soviet submarine movements. Those missile subs were *very* close to the coast. Sure, each of them had a US Navy attack boat following. Sure, if they launched a missile they would be blasted. But if they timed it right, each of those old Delta boats could get off at least two missiles, and if the close-range defences did not work the East Coast and California would be an inferno.

As he watched the screen, he saw two of the submarine symbols turn red. It really was happening. Two more. Then along the length of the Trans Siberian Railway the computer map of the USSR looked like it was erupting in measles. Each red spot indicated a missile launch. Now, the future of America was in the hands of the computers and the battle systems developed over the last 20 years. *Now* we'll see, he thought . . .

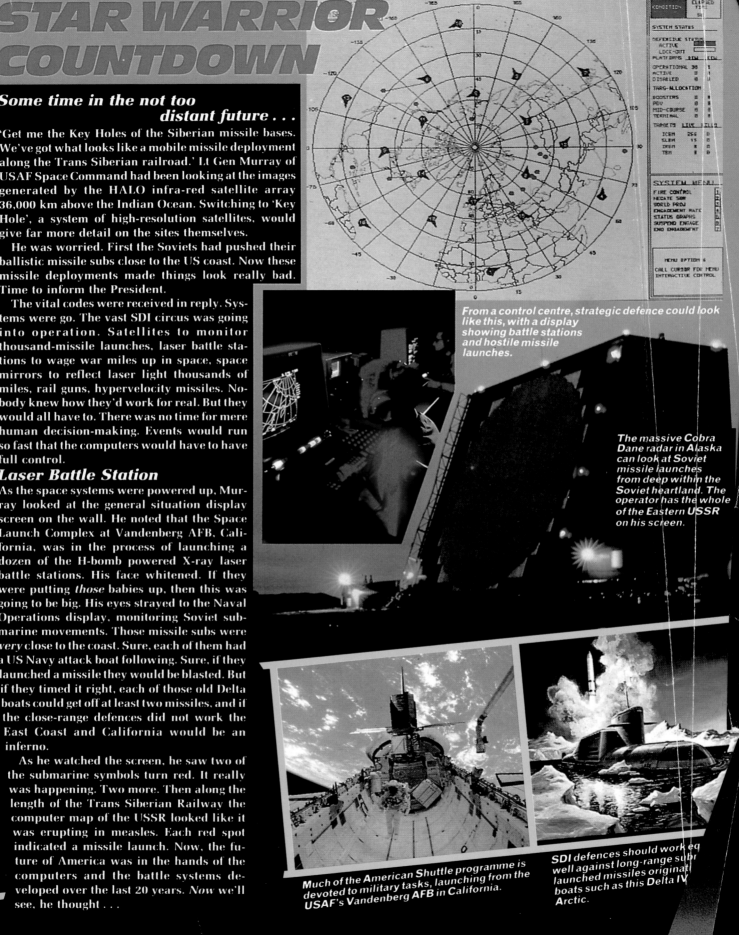

From a control centre, strategic defence could look like this, with a display showing battle stations and hostile missile launches.

The massive Cobra Dane radar in Alaska can look at Soviet missile launches from deep within the Soviet heartland. The operator has the whole of the Eastern USSR on his screen.

Much of the American Shuttle programme is devoted to military tasks, launching from the USAF's Vandenberg AFB in California.

SDI defences should work eq well against long-range subr launched missiles originat boats such as this Delta IV Arctic.

the bullet

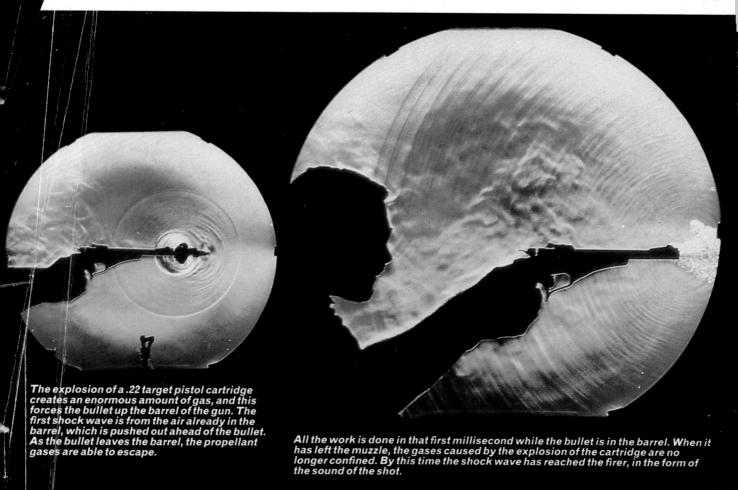

The explosion of a .22 target pistol cartridge creates an enormous amount of gas, and this forces the bullet up the barrel of the gun. The first shock wave is from the air already in the barrel, which is pushed out ahead of the bullet. As the bullet leaves the barrel, the propellant gases are able to escape.

All the work is done in that first millisecond while the bullet is in the barrel. When it has left the muzzle, the gases caused by the explosion of the cartridge are no longer confined. By this time the shock wave has reached the firer, in the form of the sound of the shot.

« Modern military rifle bullets spin at rates as high as 6,000 revolutions per second »

Most people think of a bullet as simply a lump of lead, but there is a lot more to it than that. The bullet has to be designed with definite objectives; the designer has to think about what it is going to do and then ensure that it does it efficiently. A revolver bullet for the cowboy's favourite Colt .45 may, indeed, be simply a slug of lead with a round nose, but the armour-piercing incendiary bullet for a 12.7 mm heavy machine gun is quite a complicated device.

Loading the lethal lead

In the first place the bullet has to be loaded into the chamber of the barrel from which it is to be fired. This may be a simple matter of using the fingers to load the complete cartridge – case and bullet – into the cylinder of a revolver; on the other hand the cartridge may be intended for use in an automatic weapon firing perhaps 10 or 20 times a second, in which case it must be fed into the chamber mechanically at very high speed. Then, the bullet has to be very firmly fixed to the cartridge case, otherwise it might jerk loose and jam the weapon.

Explosive charge

Once inside the chamber the cartridge is fired, the explosive powder in the case being ignited by a cap. This generates gas that, confined in the small space inside the case, develops a very high pressure. The cartridge case is securely blocked at its rear end and cannot move; the only way the pressure can escape is by forcing the bullet out of the case and through the gun barrel. This pressure, in a modern military or sporting rifle, can be as high as 4,200 kg/sq cm, and the bullet is driven into the barrel with such force that it moulds itself to the shape of the rifling.

Spinning for stability

The rifling consists of a number of spiral grooves cut into the interior surface of the barrel, and their object is to grip the bullet and make it spin. Spinning the bullet around its centre gives it stability; the reasons are complex but can be compared to the gyroscope which, once spinning, resists any attempt to alter its position. Similarly, a spinning bullet resists attempts to tumble in flight or be violently swept aside by wind currents. But the amount of spin given to the bullet depends upon the bullet's

Orbital laser relay mirror
The laser beams generated by extremely powerful ground stations could be collected, directed and focussed on missiles while they are still in the boost phase.

BOOST PHASE: THE FIRST 5 MINUTES

Soviet missiles are at their most vulnerable during the boost phase, when lifting off and accelerating towards their ballistic trajectories. At this stage they are comparatively large and slow, can easily be seen with their fiery plumes, and have not yet deployed their multiple warheads and decoys. It is estimated that the computer software needed during this phase will require 10,000,000 lines of perfectly-written instruction (by comparison, a Space Shuttle launch uses about 90,000 lines).

Launch
The USSR is more dependent upon land-based missiles than the USA, with about 2,000 missiles and 20,000 warheads. Their missile sites are much more widely dispersed than in America.

《《 *The Rail Gun will fire projectiles 60 times a second at speeds of 100 km per second* 》》

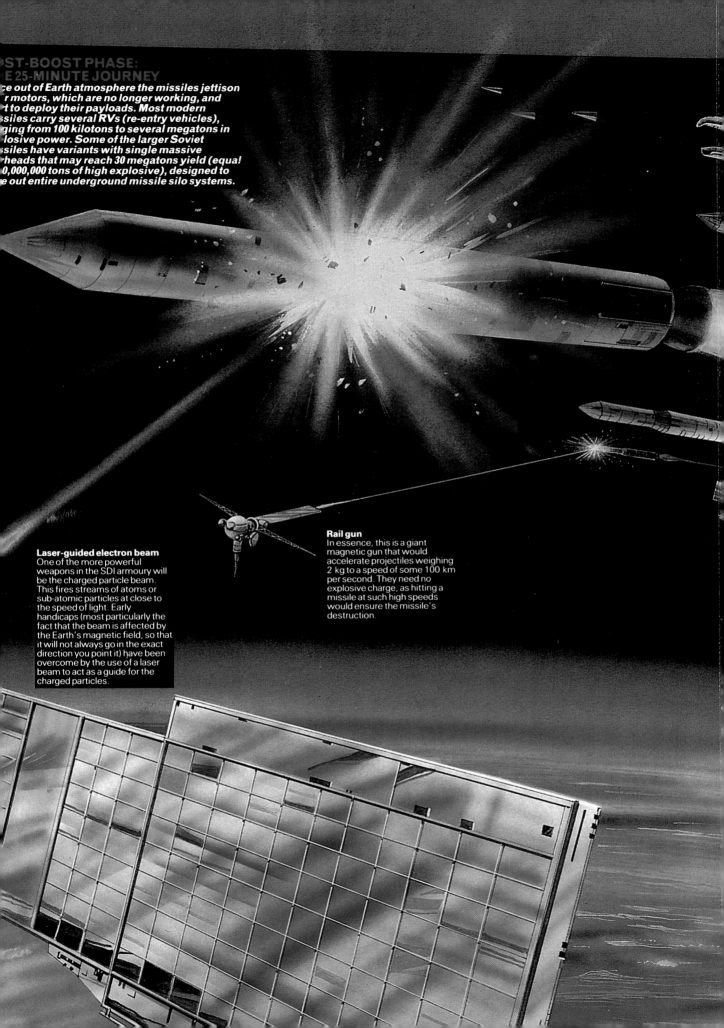

ce out of Earth atmosphere the missiles jettison
r motors, which are no longer working, and
t to deploy their payloads. Most modern
siles carry several RVs (re-entry vehicles),
ging from 100 kilotons to several megatons in
losive power. Some of the larger Soviet
siles have variants with single massive
heads that may reach 30 megatons yield (equal
0,000,000 tons of high explosive), designed to
e out entire underground missile silo systems.

Laser-guided electron beam
One of the more powerful
weapons in the SDI armoury will
be the charged particle beam.
This fires streams of atoms or
sub-atomic particles at close to
the speed of light. Early
handicaps (most particularly the
fact that the beam is affected by
the Earth's magnetic field, so that
it will not always go in the exact
direction you point it) have been
overcome by the use of a laser
beam to act as a guide for the
charged particles.

Rail gun
In essence, this is a giant
magnetic gun that would
accelerate projectiles weighing
2 kg to a speed of some 100 km
per second. They need no
explosive charge, as hitting a
missile at such high speeds
would ensure the missile's
destruction.

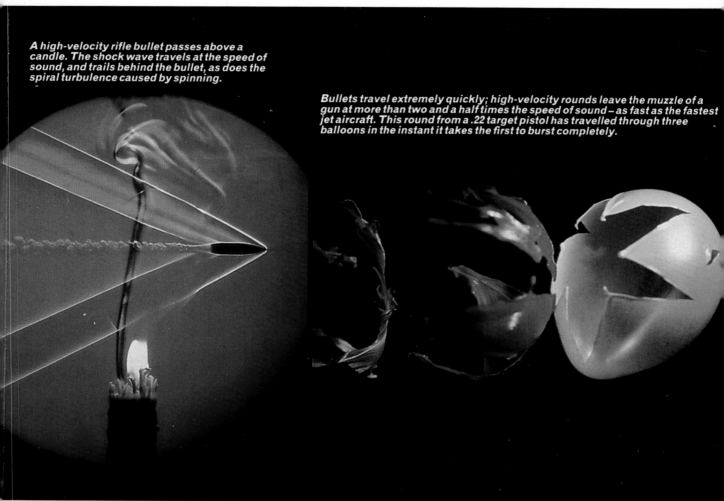

A high-velocity rifle bullet passes above a candle. The shock wave travels at the speed of sound, and trails behind the bullet, as does the spiral turbulence caused by spinning.

Bullets travel extremely quickly; high-velocity rounds leave the muzzle of a gun at more than two and a half times the speed of sound – as fast as the fastest jet aircraft. This round from a .22 target pistol has travelled through three balloons in the instant it takes the first to burst completely.

length; the shorter it is, the less spin it requires in order to be stable.

Picking up speed

The bullet, driven into the rifling, takes up its required rate of spin fairly gradually, as it picks up speed. While the bullet is moving up the barrel the propelling charge is still exploding, the pressure is building up, and the bullet is accelerating. Even when the explosion has stopped, the gases continue to expand and although the rate of acceleration decreases, the bullet is still picking up speed. Ideally the explosion of the powder should be complete by the time the bullet is about two-thirds of the way up the barrel, allowing the expansion to drive it the rest of the way and also allowing it to settle down to a steady rate of speed. The designer ensures that the 'all-burnt' position is in the right place by careful selection of the type

of powder and also of the size and shape of the individual grains of powder in the cartridge.

Finally the bullet leaves the muzzle of the gun. This is the last point at which the gas has any power, and from that moment on the bullet is 'free-wheeling' through the air, gradually losing speed as the pressure of the air resists its flight and the force of gravity tries to pull it to the ground. As a result, the path the bullet takes – its 'trajectory' – is not a straight line but a gradual curve. If the weapon is perfectly horizontal the bullet will travel horizontally for a few metres but thereafter will begin to fall to the ground.

Sighting compensation

For this reason, the sights of a weapon, particularly if it is expected to hit long-range targets, have to be adjusted so that the weapon is actual-

ly pointing slightly upwards. T compensates for the tendency of bullet to fall and ensures that it me the target at the required distar from the muzzle.

Enormous energy

By the time the bullet reaches target it has lost some of its spe Even so, there is still enough ener contained in the bullet to do dama The energy is a product of the bulle weight and speed; it can be measur in various ways, but the tradition system refers to the energy in fo pounds; thus, 10 foot-pounds mea the force of a 10-pound weight falli one foot. Our military rifle bullet h a muzzle energy of 2585 ft/lbs; at 5 metres range it is 951 foot-pounds. blow of 70 foot-pounds is sufficient knock over a man, so the 7.62 m bullet has ample energy at 500 met to do plenty of damage.

A bullet may be small, but the energy it contains is considerable. Energy is dependent upon two things: the mass of an object, and the speed at which it moves. Bullets move so fast that their impact is devastating – as shown by the effect of this one on a chocolate bar, which could just as easily be human flesh.

HE BULLET FAMILY

llets come in a wide variety of shapes and sizes, depending upon the jobs that ey have to do; for instance, a .22 rifle bullet used indoors for short-range target ooting differs greatly from a machine-gun bullet used in combat.

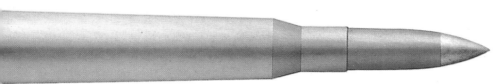

Browning .50-inch (12.7mm) machine-gun bullet has been in service for 50 years. Like most bullets in use y it has a copper jacket, but also has a steel core and tip to allow some armour-piercing ability.

baton round is very different to the ordinary bullet; fired from special riot guns, it is designed to bruise and ck, although at short range it can be as lethal as any other weapon.

The simple **.22-in bullet** is the same diameter as the NATO 5.56mm round.

A standard round worldwide is the **5.56×45,** also known as the .223 Armalite.

The **7.62mm×51** has been NATO standard for almost 30 years.

ICEBERG

On the calm, dark night of 14 April 1912, the passenger liner SS Titanic ploughed majestically through the cold seas of the North Atlantic on her maiden voyage between Britain and the United States. She was the largest ship in the world, the most luxurious – and, so it was claimed, unsinkable. But suddenly she grazed along the side of an iceberg floating silently in a peaceful sea, and pulled the rivets from her steel hull for a quarter of the length of her flank. Slowly and tragically, she sank and perished, with the loss of 1,503 lives. The iceberg that sank the liner was a tiny member of a giant population of floating ice mountains.

ARCTIC ICEBERGS

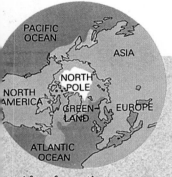

After formation, an Arctic iceberg moves out into the open sea. It remains in the Baffin Bay area for up to two years, melting and 'calving' – forming smaller icebergs – and has lost 90 per cent of its bulk by the time it reaches the coast of Newfoundland and the Great Banks of the North Atlantic. Here the warmer waters of the Gulf Stream meet the freezing Labrador current, and the iceberg has only a few days to live.

Iceberg population
At any one time, there are like to be over 50,000 icebergs in the world!

Assorted sizes
Arctic icebergs vary enormously in size, from 'growlers', about the size of a piano, to over 120 metres in height – as high as a 36-storey office block. The largest northern iceberg ever sighted was estimated to be 65 square kilometres in size – big enough to hold 100,000 football pitches!

Jagged shape
Unlike their Antarctic counterparts, northern icebergs are usually irregular in shape, having chipped off rugged glaciers, and rise steeply from the water to sharp peaks.

Dark ice
Some Arctic icebergs are almost black from vast amounts of mud and stone embedded in the ice, from when they were glaciers carving through the valleys of Greenland or Alaska.

Hunting the iceberg
Icebergs are a constant hazard to shipping, but their detection and destruction is difficult; twenty 1,000-lb bombs chipped away only about a fifth of a 250,000-tonne berg, and one frightening suggestion has been that nuclear bombs could be used! The wholesale destruction of icebergs would have a disastrous effect on the world's climate.

Ice from long ago
The ice in all the world's icebergs today is made from snow that fell up to 100,000 years ago – long before modern man inhabited Europe.

Smooth giants
South Polar icebergs are usually much larger than Arctic ones, and are more regular in shape, as the ice sheets from which they are formed are smoothed by long exposure to sea and weather.

'Fast ice'
Sometimes smaller Antarctic icebergs get stuck fast in shallow water where they provide an anchorage for drifting sea or pack ice.

《 The largest iceberg ever seen was bigger than the whole of Belgium, measuring 335 km long and 96 km wide. It was discovered in the South Pacific in 1956, accompanied by two other giants 》

FORMATION OF AN ARCTIC ICEBERG

GLACIER

Arctic icebergs are usually irregular in shape, having split from glaciers at the water's edge.
The vast glaciers flow slowly down the coastal valleys to the sea. As the ice pushes out into the ocean, huge chunks of ice shear off to form icebergs, which are then pushed slowly south by prevailing currents and wind. At this time the ice itself may be over a thousand years old, but it will survive as an iceberg for only a year or two.

ANTARCTIC ICEBERGS

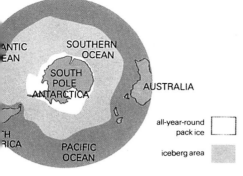

Antarctica's ice sheets account for over 90 per cent of all the ice on earth and, if they remain close to the Pole, the icebergs can survive for up to 10 years. The origin of huge tabular icebergs in this region was proved by Sir James Ross in 1841 when he penetrated to the edge of an ice shelf which now bears his name.

WHEN THE ICE MELTS

If all the ice in the world melted, the sea level in the oceans would rise by nearly 40 metres. Virtually all the world's cities would be under water, and a huge amount of the world's existing land mass would be lost.

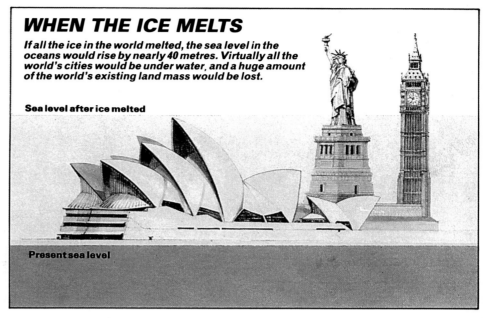

Sea level after ice melted

Present sea level

old and white
ntarctic bergs are a glacial white in colour, and along their sides you can see stripes made by layers of frozen snow; each layer represents a year's snowfall.

Both ends of the world
Icebergs are formed at both the north and south poles, and are fragments of ice sheets and glaciers originating on the mainland. As they are made from fresh water they are less dense than the sea and therefore float, with the greater part of their bulk – often 90 per cent, but sometimes only about 70 per cent in the case of Arctic bergs – below the waterline.

The rugged surface of a northern berg is broken up by cracks and fissures, which the wind erodes into dramatic caves and arches. Larger icebergs sometimes 'calve' as portions break off to form separate, smaller bergs. This can cause the centre of gravity in the remaining large mass of ice to shift so far that, in spectacular and thunderous fashion, it turns right over in the water.

FORMATION OF AN ANTARCTIC ICEBERG

ICE SHEET

WARMER WATER: ICEBERGS BREAK OFF ← 800 kilometres → COLDER WATER

In the Antarctic, sheet ice about 1200 m thick spreads out from the land, becoming shallower as it meets warmer water.
Finally, often as much as 800 km from land, the melting process causes icebergs about 200 m deep to break away. They can change shape as they float; trapped in pack ice, they often have to travel through narrow channels and, as their bulk is squeezed in all directions, their thickness may increase dramatically.

Laser Eye Surgery

When eye specialists started to use lasers in operations, a new era of surgery began. Lasers work with absolute precision, are totally germ-free, and allow surgeons to perform operations that were impossible with conventional instruments.

A surgical laser is essentially a finely controlled burner – one that uses the heat generated by light to do its work. The laser can be used like an ordinary scalpel to cut tissue, but it can also spread its beam and fuse tissues together instead of separating them. So, while lasers are now used instead of blades to cut growths from the interior of the eye, they can also repair damage at one stroke.

'This won't hurt a bit.' A laser surgeon approaches with intense red laser light emanating from the end of his operating torch. Just a microsecond of intense superlight precisely aimed at a damaged blood vessel at the back of the eye will seal it – a task previously carried out in full-scale surgery.

THE TARGET AREA

Laser light comes in various colours, determined by the characteristics of the gases and crystals used in their construction. Each colour can be used in a different way, to carry out various tasks on separate parts of the eye or even to trigger off the effects of special drugs. They must be used with great care; it takes only one tenth of a second to cause serious damage. Depth of surgery can be altered by the power and size of the burn, and width by increasing the exposure time.

Cornea
The window of the eye, allowing light rays to pass through

Lens
Directly behind the iris this changes shape in order to focus light on to the back of the eye

« Lasers cut tissues by emitting very high wavelengths of light energy that cause mini explosions or shock waves »

This real eye has been cut in half to show its components.

At the back of each eye is a layer of light-sensitive cells that collects the image coming through the lens at the front. This is called the retina, and it transmits visual information to the brain. This photograph is of a normal retina; often the fine blood vessels get blocked or burst and lasers can be used to repair them.

Retina
This layer of cells at the back of the eye is sensitive to light; from it the optic nerve transmits messages to the brain

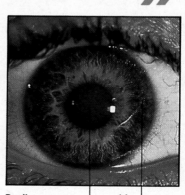

Pupil
The dark hole in the middle of the iris varies in size according to light conditions

Iris
This coloured ring of muscle expands and contracts to regulate the amount of light entering the eye

GENERATING LASER LIGHT

Laser surgeons have to precisely aim their fine point of light to hit the microscopic areas needing surgery. Lasers are used for 'sticking together' layers of tissue that have been torn, destroying abnormal growth or sealing off blood vessels.

This is one of the devices used in laser eye surgery. Another job for the laser is to burn a precisely measured drainage hole in the irises of glaucoma patients.

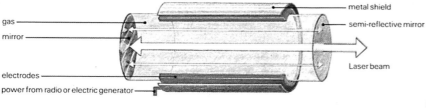

gas —
mirror —
electrodes —
power from radio or electric generator —
metal shield
semi-reflective mirror
Laser beam

Energy pumped into low-density gas causes it to emit light. Bounced back and forth between two mirrors, the light is focussed into an intensely strong beam that is able to escape through the semi-reflecting mirror. This is the laser beam.

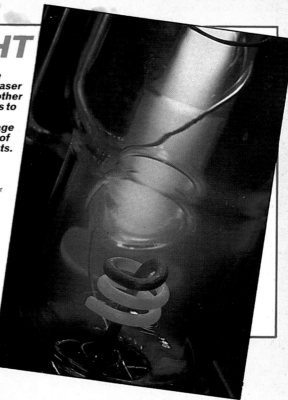

‹‹ In just one tenth of a second, serious burns and irreparable damage can be caused even by a laser as small as the ones used in eye surgery ››

Above: A fibre-optic cable directs lower power red argon laser light on to light-sensitive drugs injected into the eye of a patient. The light will activate the drug which will, in turn, destroy cancer tumours within his eyeball.

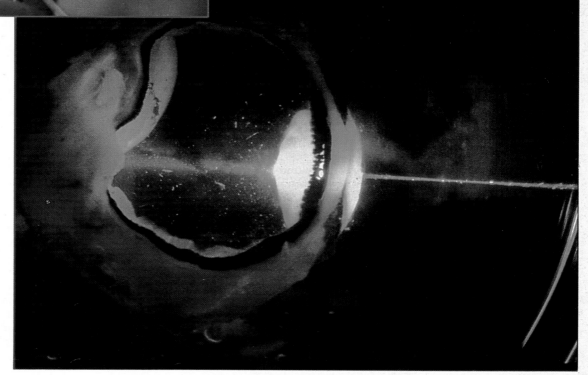

Right: A burst of blue argon laser beam travels through the cornea and lens of a cut-away eyeball and on to the retina, where it will heat-seal damaged blood vessels.

The Lambeth Body

When a dead body is found in suspicious circumstances, the first expert to be called in by the police is usually the forensic pathologist. He is a highly specialised surgeon whose job it is to determine how and when death occurred; whether it was due to suicide, accident or murder; and what weapon, if any, was used. All these objectives were spectacularly achieved by Dr Keith Simpson in the summer of 1942, when he investigated the case that the newspapers termed the 'Body in the Vault' mystery.

It was wartime, and for two years London had been raided almost nightly by German bombers. Hundreds of civilians had been killed, and Dr Simpson had what must have seemed the unending task of examining their remains in his laboratory at Guy's Hospital.

Workmen had been clearing the rubble of an old chapel in Lambeth, South London, bombed a few weeks previously. In the chapel vault they had found human remains, which on the face of it could be either those of a person who had been buried normally and exposed by the bomb blast, or killed during the air-raid which destroyed the building.

Laid out on Dr Simpson's autopsy table, the fragile pieces seemed to present little enough for the pathologist to go on. There was a skull, with scraps of hair clinging to the scalp, but no lower jaw. The skull had been separated from what remained of the torso, which consisted of an almost bare rib-cage, neck and backbone, though the abdomen and pelvis retained mummified portions of dried flesh and internal organs. The lower arms and the lower legs were missing.

Fire and quicklime

The doctor peered carefully through a hand-held magnifying glass. Part of the body was blackened, as if it had been subjected to fire. That was reasonable enough, in view of the circumstances, but it was also coated in quicklime, a substance still used by gardeners as a pesticide and, long ago, at the time of the Great Plague in the burial of bodies so as to kill the plague germs. According to popular theory a body buried in quicklime decomposes more rapidly; in fact, the substance tends to act as a

preservative, as had happened in this case.

Dr Simpson's suspicions were aroused. Still using the hand-lens, he examined the base of the skull and the ends of the upper arms and legs. No bomb blast had caused these injuries, for the marks of a saw showed clearly. There were also saw marks on the skull where the lower jaw had been removed. Clearly someone had cut up the corpse, and had probably tried to burn it before burying it in quicklime in an effort to conceal the evidence.

Now began a thorough examination. An expert can 'sex' a skeleton very quickly, principally by the size and shape of the skull and pelvic bones, and this was indubitably the

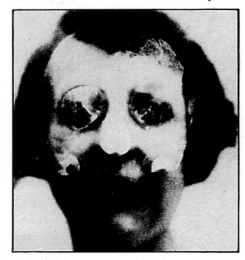

A photograph of the suspected victim was superimposed on the skull, and fitted perfectly: another link in the chain of evidence. By a grim irony, it was Dr Keith Simpson who performed the legally-required autopsy on the murderer Harry Dobkin's body in the shadow of the gallows, an hour after the sentence had been carried out. Simpson always maintained that it was the Dobkin case that pleased him most.

remains of a woman, for if further proof were needed Dr Simpson found the womb, mummified but intact, in the abdomen. Inside the womb was a large fibroid tumour, a form of non-malignant cancer, which must have caused a great deal of discomfort.

Further examination, particularly of the teeth, the skull sutures (fine joins on the cranium, where the three principal bones of the skull meet), and the laminal spurs (bony 'bumps') on the spine, gave Dr Simpson a fair indication of the woman's age. And by measuring the remaining skeleton and making a scientific assessment of

the probable length of the missing lower legs and feet, he was able to arrive at an accurate idea of her height. Meanwhile, the scrap of hair had been washed in the laboratory; it was brown, streaked with grey.

Finally, Dr Simpson had a great stroke of luck. Part of the throat had been preserved by the action of the quicklime, and on dissecting it he found that the tissues had been badly bruised and that the hyoid bone, a delicate little horseshoe-shaped bone just above the thyroid, had been broken. There is only one common cause for such a fracture: strangulation.

So from these pathetic scraps of human tissue, Dr Simpson had a picture to give the police. The dead woman was aged between 40 and 50, between five feet and five feet one inch tall, with greying brown hair and a painful tumour in her womb.

Caretaker questioned

Scotland Yard were delighted. While Dr Simpson made his detailed examination, they had been conducting their own enquiries, and had discovered that the last caretaker at the chapel had been a 49-year-old man named Harry Dobkin. Furthermore, Dobkin's wife Rachel was 47, five feet one inch tall, with grey-brown hair, and the records of two London hospitals showed that she had twice refused to have an operation for a womb tumour.

She was separated from her husband, who had served a term of imprisonment for non-payment of legal maintenance to her. And Rachel had been reported missing 15 months previously.

Just to clinch matters, police found Mrs Dobkin's dentist, who was able to produce his records of work done on her upper jaw; they matched the remaining teeth perfectly. Then, using a technique perfected 15 years previously, a photograph of Mrs Dobkin was superimposed on to a photograph taken of the 'chapel' skull. These, too, were a perfect match.

Harry Dobkin was traced by police and, presented with the forensic evidence, broke down and confessed. He had fallen behind with his maintenance again, and rather than risk a further prison sentence had killed Rachel and disposed of her beneath the chapel vaults, taking a gamble that she would not be found – a gamble which might well have come off.

SUPERMAN UNDER THE NORTH SEA

You're very strangely dressed, 600 km from dry land, in a frail-looking cage that's swinging from a winch above one of the most hostile tracts of open water on Earth. A chill wind is whipping up four-metre waves that are probably no warmer than 5°C. Slowly you sink into the water, buffeted for a while in the splash zone – the six or seven metres of turbulence near the surface – and, as you go down, the light fades around you. This isn't a bad dream. Another day's work has begun in one of the world's most dangerous jobs: diving in the North Sea.

There is almost nothing that's safe or pleasant about a North Sea diver's work. If you're lowered through the water at much more than 17 metres per minute, you can suffer convulsions. Come back to the surface too fast and your blood will literally boil, as compressed gases in your body expand faster than human flesh can stand.

At any time your life support systems and safety line can be ripped away by a sudden movement of the ship above you, or get fouled round equipment or an underwater structure. You can drown, suffer carbon

Diving under the North Sea is not easy. Much of the diver's work is fairly straightforward – welding, repairing pipes, carrying out heavy industrial cleaning – but he is in a cold, hostile environment where the smallest mistake can be fatal.

« *In the 1970s, one diver returned to the surface so fast that the decompressed gases blew his body to pieces* » dioxide poisoning, freeze to death, or shear off a limb with a high-pressure water gun or thermic lance.

All these hazards come as a bonus on top of working almost blind in deep, dark, dirty water, having your diving suit regularly infested with tiny sea-lice, and spending weeks at a time away from wife, family or girl-friend, in cramped quarters aboard a none too luxurious diving vessel where alcohol is absolutely prohibited.

Dirty work

The oil and gas platforms in the North Sea could not, however, function without the 2,000 or so men who are qualified to dive in those treacherous waters. Building the North Sea installations was an achievement as complex and difficult as putting men into space, but it was essentially an adventure in engineering. Once the rigs were in place, skilled, tough and resourceful human beings were needed for lighter underwater construction work and to inspect the submarine structures and keep them safe.

« Divers breathe a mix of oxygen and helium, since pressurised nitrogen (the normal inert component of air) acts like pure alcohol! »

The diver is very careful when he enters the water, as his life depends on the umbilical line that gives him his air supply. To make sure that no undue stress is placed on it, he is normally lowered into the water on a platform. This sometimes goes all the way down with him, providing a handy area to work from and to store tools and equipment.

PREPARING FOR THE DIVE

As a diver, you have to function under very difficult conditions. Your life depends on your equipment, so the compulsory pre-dive checks have to be thorough.

REMEMBER: A DIVER MUST ALWAYS CHECK HIS OWN EQUIPMENT. *NEVER* ASSUME THAT A PIECE OF EQUIPMENT IS SAFE UNTIL YOU HAVE CHECKED IT YOURSELF, EVEN WHEN ANOTHER MEMBER OF THE TEAM HAS THE JOB OF CHECKING AND PREPARING IT. AFTER ALL, IT'S YOUR LIFE THAT DEPENDS ON IT!

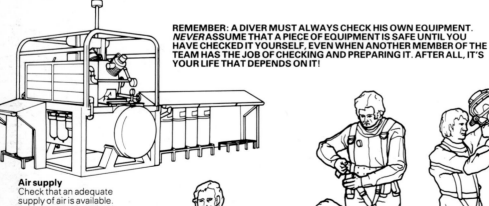

Rubber and terylene braided air hose

Communication cable

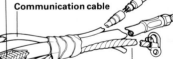

Safety line

Surface-monitored depth gauge (pneumofathometer)

Umbilical
Check for any damage and ensure that the air hose has been pressure-tested. Make sure that the air hose will not have to bear any of your weight. Check the connections of the communication line, safety line and surface depth gauge.

Ready to dive
You are now ready to enter the water. If you are not using a winch platform, a ladder entry must be *cautiously* made. Do not snag your umbilical, and make sure you are not snatched off the ladder by a swell or large wave.

Air supply
Check that an adequate supply of air is available. Any compressor or bank of high-pressure tanks must be able to produce an air flow sufficient for the depths to be dived at.

Reserve air supply
Check that the reserve cylinder is full enough to get you to the surface in an emergency. Check it for damage; check seals and valves, air hoses and harness.

Clothing
Check diving boots for excessive wear. Check diving suit for obvious damage, and check the fit – if there are creases, they will cause a condition known as 'suit squeeze' under pressure, pinching the skin in the fold area.

Final check
Before putting on your helmet, check fit of weight belt and check the emergency release. Check that the helmet air supply is locked on and then put on your helmet. Check that you can reach the emergency bottle manifold. Check that the umbilical and the helmet are secure.

THE DEEP-DIVING TEAM

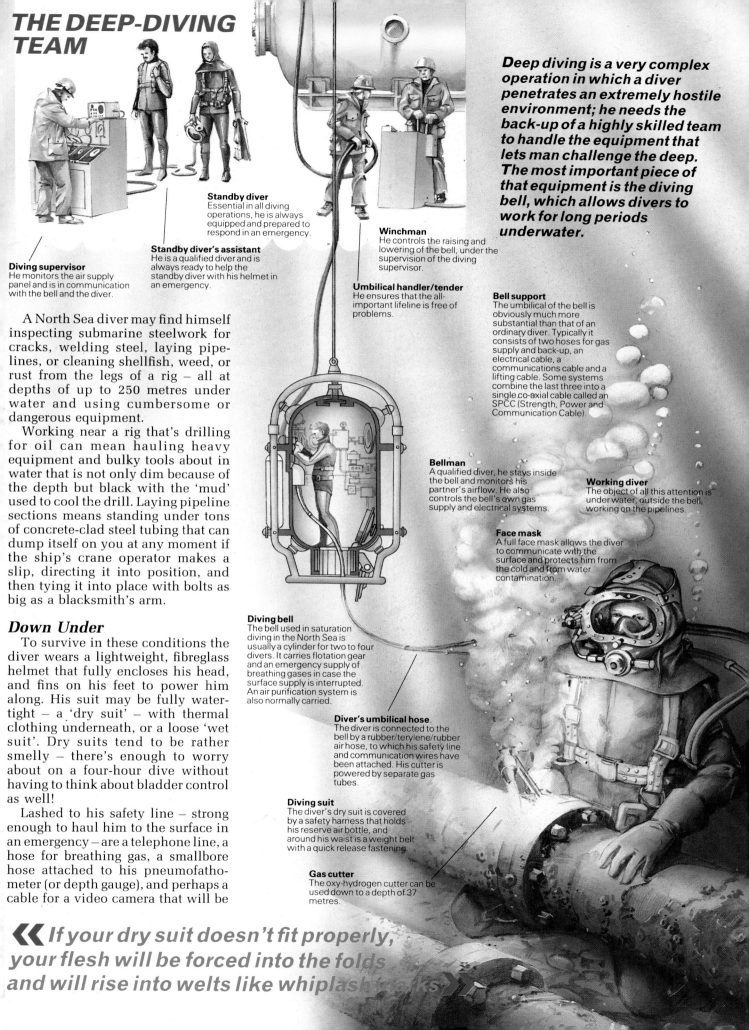

Deep diving is a very complex operation in which a diver penetrates an extremely hostile environment; he needs the back-up of a highly skilled team to handle the equipment that lets man challenge the deep. The most important piece of that equipment is the diving bell, which allows divers to work for long periods underwater.

Diving supervisor
He monitors the air supply panel and is in communication with the bell and the diver.

Standby diver
Essential in all diving operations, he is always equipped and prepared to respond in an emergency.

Standby diver's assistant
He is a qualified diver and is always ready to help the standby diver with his helmet in an emergency.

Winchman
He controls the raising and lowering of the bell, under the supervision of the diving supervisor.

Umbilical handler/tender
He ensures that the all-important lifeline is free of problems.

Bell support
The umbilical of the bell is obviously much more substantial than that of an ordinary diver. Typically it consists of two hoses for gas supply and back-up, an electrical cable, a communications cable and a lifting cable. Some systems combine the last three into a single co-axial cable called an SPCC (Strength, Power and Communication Cable).

Bellman
A qualified diver, he stays inside the bell and monitors his partner's airflow. He also controls the bell's own gas supply and electrical systems.

Working diver
The object of all this attention is under water, outside the bell, working on the pipelines.

Face mask
A full face mask allows the diver to communicate with the surface and protects him from the cold and from water contamination.

Diving bell
The bell used in saturation diving in the North Sea is usually a cylinder for two to four divers. It carries flotation gear and an emergency supply of breathing gases in case the surface supply is interrupted. An air purification system is also normally carried.

Diver's umbilical hose
The diver is connected to the bell by a rubber/terylene/rubber air hose, to which his safety line and communication wires have been attached. His cutter is powered by separate gas tubes.

Diving suit
The diver's dry suit is covered by a safety harness that holds his reserve air bottle, and around his waist is a weight belt with a quick release fastening.

Gas cutter
The oxy-hydrogen cutter can be used down to a depth of 37 metres.

A North Sea diver may find himself inspecting submarine steelwork for cracks, welding steel, laying pipelines, or cleaning shellfish, weed, or rust from the legs of a rig – all at depths of up to 250 metres under water and using cumbersome or dangerous equipment.

Working near a rig that's drilling for oil can mean hauling heavy equipment and bulky tools about in water that is not only dim because of the depth but black with the 'mud' used to cool the drill. Laying pipeline sections means standing under tons of concrete-clad steel tubing that can dump itself on you at any moment if the ship's crane operator makes a slip, directing it into position, and then tying it into place with bolts as big as a blacksmith's arm.

Down Under

To survive in these conditions the diver wears a lightweight, fibreglass helmet that fully encloses his head, and fins on his feet to power him along. His suit may be fully watertight – a 'dry suit' – with thermal clothing underneath, or a loose 'wet suit'. Dry suits tend to be rather smelly – there's enough to worry about on a four-hour dive without having to think about bladder control as well!

Lashed to his safety line – strong enough to haul him to the surface in an emergency – are a telephone line, a hose for breathing gas, a smallbore hose attached to his pneumofathometer (or depth gauge), and perhaps a cable for a video camera that will be

《《 *If your dry suit doesn't fit properly, your flesh will be forced into the folds and will rise into welts like whiplash marks*

SAFE WORKING PRACTICE

Water is not Man's natural environment – mistakes can be fatal. As a diver, you HAVE to be safety-conscious, otherwise you're dead.

Descending line
If you are not descending on a platform you should follow a line, adjusting your air supply and equalising the pressure in your ears and sinuses as you go. Your rate of descent should never be more than 30 metres per minute.

This diver is checking a pipeline for cracks and faults using an ultrasonic probe. This emits sound waves, which are reflected back from the inside surface of the pipeline, and a change of pitch indicates a fracture. The diver reads the information on a meter.

Working depth
When you reach your working depth, you leave the descending line; make sure your umbilical is not fouled with other lines and, to guard against a sudden surge or pull, loop it once around your forearm.

mounted on his helmet. If the diver's in a wet suit, another hose carries hot water that circulates inside the suit and stops him from dying of exposure in the icy sea.

The diver will also be carrying his tools – ultrasonic equipment, perhaps, to test for metal fatigue in three dimensions, or a camera, bolts and wrenches, or more sophisticated items.

On deck, the diving supervisor will be monitoring his gas supply and depth, while an assistant handles the lifeline. A winch operator constantly stands by, as does a second, fully equipped diver and assistant ready as an emergency underwater back-up.

There are usually no more than two divers from any one vessel at work at

Safety line
Follow the safety line when ascending. As the water pressure decreases, suit pressure will increase, making it necessary to vent your suit. All decompression rules *must* be followed, with a maximum rate of ascent of 20 metres per minute. The number and length of decompression stops on the way up will depend upon how long your dive has lasted and at what depth.

Care of the umbilical
Remember, this is your lifeline. If possible, pass it *over* obstructions such as rocks or pieces of wreckage to avoid fouling it.

Reaching an obstacle
When passing an obstruction always return on the same side to avoid snagging your line.

Lines and moorings
Take special care when working near these, particularly when they are under strain. Do not pass beneath them, and beware of those that have been in place long enough to collect razor-sharp barnacles. *Never cut a line until it has been positively identified.*

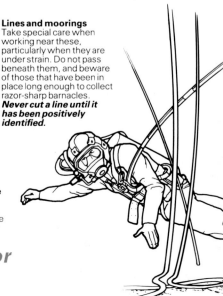

《《 *North Sea divers often have to swim in mud or quicksand – they treat it like thicker water* 》》

THE TOOLS OF THE TRADE

Many jobs that are simple above ground need very special tools when carried out under water.

Here a diver is using a high-pressure water jet on a Shell oil rig in the North Sea. The murky water makes it a difficult task.

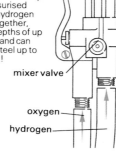

flame   nozzle

Oxy-hydrogen cutter

This tool gives a good, clean cut and is useful during demolition work – dismantling pipelines and equipment on a dry rig. The pressurised oxygen and hydrogen are ignited together, will burn at depths of up to 37 metres and can cut through steel up to 100mm thick!

mixer valve

oxygen

hydrogen

High-pressure water jet

Anything that's under water is going to get covered in marine growth, so a regular diver's job is to clean up equipment below sea level. High-pressure water jets mixed with an abrasive are very effective; the pressure is *very* high, up to 1.4 tonnes per square centimetre – more than 20,000 pounds per square inch!

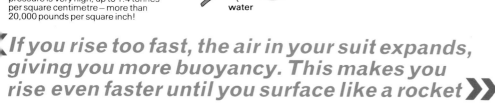

sand or aggregate   high-pressure abrasive flow

water

HAND SIGNALS

Communication under water between divers is almost exclusively by hand signal; only a few important gestures are necessary to get your point across. Most of them can be used as a question or as confirmation, depending upon whether you signal first or in response to someone else.

Go up/I am going up

Go down/I am going down (usually seen only at the start of a dive)

Stop/stay where you are

Something wrong/I have found a problem (e.g. 'I have found the damaged pipe/valve/ mooring'; NOT 'I am in trouble')

Indication of location/ Indication of person referred to in following signal

OK?/OK (question from one diver to another OR reply to that question)

I am in distress (immediate priority signal; anyone seeing it must go to assist)

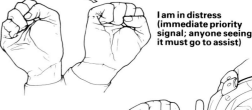

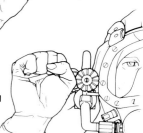

I am on reserve/I need to finish dive

If you rise too fast, the air in your suit expands, giving you more buoyancy. This makes you rise even faster until you surface like a rocket »

any one time. Usually they work in pairs, descending in a diving bell for greater safety, while each can aid the other if trouble develops.

Take a deep breath . . .

There are in fact two types of diver working in the North Sea – air divers, who breathe compressed air and work at depths up to 50 metres, and 'saturation' divers who work below 50 metres and breathe a mixture of oxygen and helium.

Deep-sea divers breathe helium because at more than 50 metres under water nitrogen – which forms 78 per cent of ordinary air and, like the diver himself, is subject to underwater pressure – has much the same effect as alcohol. No-one knows exactly why this happens, but the effects, known evocatively as 'raptures of the deep', make divers both carefree and careless – some have even been known to take off their helmets under water, with disastrous results.

Going for the burn

While contending with all the natural hazards of the North Sea, the working diver has to handle tools that are themselves lethal.

The most unnerving is the thermic

lance. The lance is part of a high-power electrical circuit. The diver latches a power line from the lance on to the metal he's going to cut. Next he turns on the electricity, which makes the nozzle itself live. Then he puts the nozzle close enough to the metal to complete the circuit. The result is a giant spark as the electricity arcs through the water. This huge surge of power ignites the pressurised oxygen.

No less dangerous is the high-pressure water jet, which blasts out a mixture of grit and water at pressures up to 850 kg/sq cm – enough not only to tear shellfish from their moorings but to crush a diver's leg to pulp, bones and all, as well.

All's well . . .

North Sea divers pay for their own training – which takes months, and costs thousands of pounds – and have to pass a stringent medical every year. They risk their lives every day and live in grim conditions. Their tenacity, professionalism and sheer courage have made Britain's venture into the North Sea the success it is, and they probably deserve even higher rewards than those they now earn.

You wake suddenly from a sound sleep. What was that noise? Could it be burglars? As you creep towards the kitchen you hear scurrying and the rattle of claws. You snap the light on, and after a frozen moment a brown rat screams at you while several more leap away from the remains of your supper. In an instant you are alone. When you investigate, you find a hole gnawed through the plastic joint connecting the downstairs lavatory with the drains; they have come up through the sewers. Cunning creatures, rats.

Mankind and rats have been inseparable since the dawn of civilisation. Two species of rat in particular have come virtually to depend on human beings for their survival: the Black rat (also known as the Ship rat or Roof rat), and the Brown rat (also called the Common rat or Norway rat).

Rambling rodent

The Black rat came to Britain from central Asia or northern India, spreading across Europe in the 11th and 12th centuries and bringing bubonic plague in its wake. Fleas from the Black rat spread the disease, which wiped out a third of the population of Europe in the 1340s. Then it sailed across the world with the European navigators – hence the name Ship rat.

The Brown rat came from central Asia too, but reached Europe only in the 18th century on ships trading with Russia. British ships carried it across the globe. The Brown rat is now found everywhere from old whaling stations in the Antarctic to Pacific Islands and the Canadian prairies – from where it is still spreading northward.

Country comforts, city slickers

In the country, rats live outdoors in spring and summer, when food is plentiful. In the autumn, as crops are harvested, they follow their food supply indoors to grain silos and barns, and spend the winter there.

Life's harder in the city for a rat. While there's plenty of rubbish to forage in, rats also survive by eating axle grease dropped by trains, and will strip electric wires bare to get at the lubricant on the wire. They need plenty

Fast breeders
A pair of rats could have 420 descendants in a year. In 10 years this could mean 48,319,698,843,030,344,720 descendants!

Rats at sea
Rats *do* leave sinking ships; they live in the bilges, which are the first places to fill with water, so the rats are often the first to know that a ship is going down.

Holy rats
There is a temple in India that has 100,000 sacred rats that are believed to be reincarnations of the souls of poets.

Farm rats
In 1901, an extermination drive on a Sussex farm killed 37,000 rats.

of water, too, and will chew their way through lead pipes to get at it.

Breeding like . . . rabbits?

Rats breed astonishingly fast. Old enough to bear young at two months, a female rat's pregnancy lasts only three weeks. In a year, a single pair can produce thousands of descendants even with each pair having only an average litter of between five and eight young – and litters of over 20 have been known. If they were all to survive, they'd soon take over the world.

But there are several reasons why we're not knee-deep in rats. In the country, they're caught and eaten by numerous predators from owls to foxes. In towns, their greatest enemy is man, who makes a considerable contribution to making a rat's life short and not very sweet – some 95 per cent of rats die before their first birthday.

The war against rats is endless, not only because they can destroy harvest but because they carry disease. Apart from the plague the list includes salmonella food poisoning, typhus, worms, and Weil's disease (a particularly damaging form of jaundice), while a rat bite will make a septic wound followed by fever.

The killing fields

Rats were long able to avoid most systematic attempts to destroy them. A rat that survives eating poisoned food will simply avoid the place it found it in future. Some rats won't touch any food that's unfamiliar.

A modern poison like Warfarin doesn't arouse a rat's suspicions, and once eaten thins out the rat's blood to such an extent that it will break through the lung walls so the rat actually drowns in its own blood. But 'super-rats' have begun to develop an immunity even to these chemicals.

Rats do have one saving grace. Their body chemistry is remarkably similar to ours, and they are invaluable in testing drugs – especially as their rapid rate of breeding allows many generations to be tested for side effects in a relatively short time. Outside the laboratory, however, they will be sending shivers up most people's spines for a long while yet.

« *It is estimated that around 100 million rats are used in experiments each year by exploitive but unappreciative humans* »

Bad habits
Rats keep themselves clean, but make a lot of mess: 300 rats produce 15,000 droppings and 3.5 litres of urine in 24 hours.

Cannibal rats
Last century, an American solved his rat problem by caging a number of rats together without food. They began to eat each other, leaving one ferocious survivor; after he was turned loose, that was the end of the problem.

The friendly rat
Socially-inclined rats are known to help one another steal eggs and dog food.

Agile rats
Though the Black rat is the best climber, the common Brown rat will have no problem climbing up a vertical wall to get into a building.

Greedy rats
It is reckoned that China's rats eat one tenth of the country's grain harvest each year – enough to feed 100 million people.

an eight-speed gearbox and an auxiliary two-speed 'splitter' attached), engine improvements mean the trend is back towards simplicity. On a big truck, operators often prefer to demonstrate their skill in handling the rugged simplicity of non-syncromesh gears to the sophistication of power-assisted gearshifters and huge numbers of ratios. In any case, on most long hauls the truck may spend as much as 75 per cent of its time running flat out in top gear – at around 35 mph for every 1000 rpm of engine speed.

Today's big rigs are vastly more sophisticated than the first heavyweights that took to the road. But underlying all the mechanical complexity of a big truck is the owner's and operator's pride – the pride that makes every truck an individual, and every trucker a folk hero. And that may never change.

Piggy back
Australian logging trailers can usually be folded and carried on the back of their tractors. This saves tyres and axle wear on the trailer.

Trailer arms
The side-arms of pole-ty trailers are extendable, f dealing with log loads of differing girth.

Sleeper box
Sleeper boxes come in many shapes and sizes. This one is the standard 90 cm (36 in) in width, but has a distinctly non-standard roof extension. These boxes usually have a bunk with luggage space underneath, and basic washing and food storage facilities. Larger, more expensive versions can be fitted out like luxurious motor homes.

Exhaust stack
The chrome-muffled exha systems are made with a bends as possible, allowi free passage of exhaust g from the engine. The back pressure is therefore low means less engine wear longer engine life.

Engine
Kenworth, like other US truck manufacturers, offers clients a wide choice of engines when they order their trucks Engine manufacturers include Cummins, Caterpillar and Detroit Diesel. Normal power output of the largest V-8s is around 450 hp (335 kW), but for hauling their very heavy loads up steep gradients loggers will often have the engines modified to give a power output of up to 600 hp (450 kW)

The interior of a modern interstate cab is designed to make the driver as comfortable as possible on his long journeys. Hydraulic seats, radio and air conditioning are all standard items these days.

ing fuel costs, the trend today is towards less power (perhaps 'only' 400 bhp) but more efficiency.

Whereas a big diesel of the late 1970s might have held a steady 2,100 rpm (very leisurely by, say, a car engine's standards), today it might only run to 1,600 rpm – reducing fuel consumption by maybe 25 per cent by this means alone.

The other major way of improving economy is through aerodynamics, although a model like the Kenworth hardly seems to recognise the problem at all. A big outfit like this may be displacing as much as 12 tonnes of air per kilometre, and the effects of smoothing out the shape and filling the turbulence-filled gaps can be considerable. What's more, a 10 per cent increase in speed from a maximum of 60 to 66 mph can cost as much as 25 per cent more in fuel consumption, so modern rigs offer a growing range of aerodynamic tweaks, and even computer-controlled monitoring of engine performance to help a driver wring the most out of his truck.

≪ **Some trucks are fitted with colour TV, video, stereo hi-fi, microwave, refrigerator, sink, chemical toilet – and even a king-size airbed** ≫

with shaped panels rivetted onto a rigid frame.

As well as the basic working cab, the creature comfort options open to an American long-haul operator are virtually limitless. Anything from simple pile carpets and air-sprung seats, plus the inevitable CB radio to keep in touch – on 40 channels – with the outside world, to a fully equipped sleeper cab with bed, air-conditioning, even cooking and washing facilities, is available at a price, which can reach as high as $170,000. Many of today's rigs are virtually working mobile homes.

Pedal to the metal

A radiator of up to 144 square inches frontal area allows the customer to specify a wide choice of diesel engines from standard units to the most powerful, with or without turbocharging (exhaust-driven supercharging, introduced by Volvo in the mid-1950s). These are available mainly from the big three engine makers, Caterpillar, Cummins or Detroit Diesel. These can include V8s of up to 18 litres, and power outputs up to 600 bhp (448 kW) but, thanks to ris-

Diesel cowboys

The choice of transmissions is just as wide but, in spite of the possibilities of large numbers of gears (say 16 with

Cab
In spite of the power and size of the tractor, the Kenworth C500 has a relatively small cab. The twin chrome-plated airhorns and the extra lights are standard fittings on Kenworths; cabs are usually of aluminium with GRP roofs, but customers can specify steel cabs.

Air filter
The massive air filters for the V-8 diesel are mounted outside the engine compartment.

Paint jobs are the pride of the trucker, and express his own ideas and individuality. The big truckmakers offer a wide range of schemes, but many trucks have original custom paint jobs.

≪ **A well-maintained truck will often travel more than half a million kilometres (310,000 miles) in five years or less.** ≫

Radiator
The enormous radiators, so typical of big American rigs, vary in size according to the power of the engine. Kenworth radiators range from 6250 square cm to 9030 square cm. The largest are designed for use with engines of up to 525 hp (390 kW).

BIG RIG

A fully-loaded refrigerated truck ploughs ever onwards through the rain. At 55 miles per hour, the giant throws up 450 litres of water from its tyres every minute!

The average fuel consumption of a highway heavy is about four miles per gallon

W ith a noise like a bad-tempered dragon clearing its throat, a shudder and two dirty gouts of smoke from its shiny upright exhaust stacks, 50 tonnes of fully loaded custom-made Kenworth truck comes to life. A long grumble from the 18-litre engine, and the first of 13 gears takes hold: the 18 metre-long rig, complete with air-conditioned living quarters, crawls out towards Interstate 10 and the first leg of an 11,000-km journey from Los Angeles to Montreal.

This rig will haul 30 tonnes of freight over perhaps 250,000 km in a year from coast to coast across the continental USA, thundering along at a more-or-less legal 60 mph, its 450-bhp (336-kW) V8 engine gulping 70 litres of gas per 100 km.

Tailor-made truckin'

To the dedicated long-haul truck operator, this Kenworth 'conventional', with its classic bonnetted nose and custom paintwork, is the hard-working equivalent of a made-to-measure suit; whatever the operator wants – horsepower, load capacity, an individual external appearance or a sumptuous cabin – the manufacturer is only too pleased to build it.

Replete with customised paintwork, smothered in chrome, and operated more often than not by an independent owner – one of 'the last of the cowboys' – the 18-wheeler big rig is a potent symbol of American freedom, individuality and, above all, sheer macho style.

Like fellow American manufacturers of the industry's big rigs, Mack, White, Peterbilt and International, Kenworth, of Seattle, Washington State, pride themselves on the number of options they can offer – to the extent, in fact, that each truck is built to the customer's precise individual specification.

The basic choice in these top-league rigs is between this traditional long-nosed tractor, or the flat-fronted, forward-control, cab-over-engine type. The cab-over may look more modern but is still marginally less popular in the vast American trucking industry. Most operators say the conventional gives a more comfortable ride with more cabin space, easier access and better frontal protection in the event of an accident. And it has the stop-at-nothing look s that are still so important to the trucker, even in today's highly competi-tive industry.

Roll, big wheels, roll

The Kenworth is based on a massively strong channel-section, ladder-type chassis – hand-built in steel and aluminium for maximum strength with weight. Every kilogram saved in construction is another kilogram added to the rig's cash-earning carrying capacity within the all-up legal limits. Weight is saved too in the cab structure, a light, strong shell made almost entirely of light alloys, with

One thing these huge vehicles lack is good aerodynamics. The massive radiator can be up to 9,000 square centimetres in frontal area, contributing to a fuel consumption of about four miles per gallon.

The United States is so huge that a truck will take several days and nights, travelling under the legal speed limit of 55 mph, to cross the country. The drivers have sleeping compartments behind their cabs; driving in two-man teams, one man can rest while the other handles the rig.

《 The USA has a staggering 35,000,000 heavy trucks licensed – or one heavy for every seven American citizens! 》

...kers are the oil that greases the wheels of ...nighty American economy. The majority of ...ht is carried by road, and the interstate ...kers, with their immense vehicles travelling ...aps 200,000 miles per year, are the kings of ...oad.

When you've got 350 hp available, you can give your mates a hand – in this case three mates. These Macks are on an interstate delivery run, piggy-backing on a cabover (cab over engine) cruise-liner to save fuel and driver costs.

This C500 Kenworth works for a logging outfit in Tasmania, hauling gum trees out of the forest along rough dirt roads to the pulp mills 160 km away.

BOGIE BUSINESS

Public roads usually have a maximum weight limit and a maximum length for trucks, so a logging truck can usually tow only a single trailer. The Australians normally carry a second trailer piggyback on the tractor unit.

On private roads, national and state regulations no longer apply, and the only limits to the load trucks can tow are engine power and safety. Thus the second trailer comes into action.

Typically, a single trailer load weighs about 23 tonnes, but with a twin trailer the same truck can carry 50 tonnes. Total all-up weight for the loaded combination is about 70 tonnes.

Braking
The stopping power of a large truck is adequate on the open road – but when a tractor is going downhill with a load of 70 or more tonnes behind, something more is required. In addition to the massive drum brakes on the wheels, trucks of this type usually have an engine braking system, as well as a retarding system built into the automatic transmission. The wheel brakes themselves usually have their own water cooling system.

Trailer
The pole-type trailer connects to a tow-hook on the tractor unit; the swivelling turntable on the forward bogie allows for easier steering. The trailer extends lengthways for carrying longer treetrunks.

Payload
Wood varies considerably in density, so a 50-tonne load of hardwood might take up considerably less space than a similar weight of young conifers. A typical load of pine might be stacked 7 metres high on the trailer.

C500
KENWORTH

Logging trucks are among the most interesting of all heavy haulers, particularly those used in the lumber industries of Canada, the USA and Australia. They mostly operate on private, unsurfaced roads in conditions that the average trucker would consider impossible, hauling enormous loads that would be illegal on the public highway. Obviously, the trucks themselves must be a little bit special.

Konishiki, the 1985 Grand Champion, displays his awesome bulk. Sheer size is useful to a fighter, but success requires more than that: skill, agility and aggressive spirit all have their place in the ring.

SU
W R E S

Two professionals co together with a groun shaking crunch during the tachai, the charge with which every bout begins. Note the widespread hands, equally ready to slap, push or grapple.

« **A sumo wrestler's main meal comprises five or six large bowls of a high-calorie stew (meat or fish, cabbage and spinach, onions, malt, tofu, sugar and soy sauce, in a rich stock) accompanied by equal amounts of rice; this noonday meal is usually accompanied by several pints of beer. He then sleeps for most of the afternoon. A past champion, Dewatage, stood 2 metres tall and weighed nearly a quarter of a tonne!** »

SUMO THROWS

There are three main winning throws in *sumo*. Slapping and thrusting moves (*Tsuki Dashi* and *Hataki Komi*) involve slapping the opponent out of the ring, and dodging and slapping him to the ground. Continuous pushing includes such moves as *Nodowa*, the throat push. Also, there are grappling techniques such as *Uchi-Gake* (the inside trip), *Kote Nage*, the forearm-grasped hip-throw, and *Utchari*, the defensive spinning throw.

TSUKI DASHI

HATAKI KOMI

If pushing and shoving doesn't work, the wrestlers grasp each other's loincloths. In this instance the nearer man has his arms inside those of his opponent, and has succeeded in lifting him bodily. If he can carry him out of the ring – a move called yuri-dashi – he will have won the bout.

First come two men dressed in the uniforms of 13th-century riding masters. Next comes a swordbearer, also in medieval garb. Behind him lumbers the incredible bulk of a 25-stone (165-kg) man. His hair is shaved off but for a topknot sprouting from the top of his head, and he is wearing only an embroidered silk apron over a silk loincloth. This is *dohyo-iri*, the entrance ceremony that announces the start of a Japanese *sumo* wrestling match.

Placating the spirits

The battle for advantage is on as soon as the fighters are in the ring – a 4.5-metre diameter coarse hemp mat atop a dirt mound. The huge wrestlers rinse their mouths with holy 'power water', scatter salt into the ring, and stamp their feet, both to purify the ring and drive away evil spirits.

At the same time, the fighters are each trying to 'psych' the other out, scowling, swaggering, or simply ignoring each other. Then, flexing their muscles, they stamp themselves into a widespread squatting posture opposite each other and begin to coordinate their breathing. The moment of *tachai* – meaning 'sudden and surprising even if expected' – is approaching.

Then the enormous wrestlers launch themselves at each other. The fight is on in earnest.

Counted out

The simple aim of a *sumo* wrestler is to force his opponent out of the ring, or throw him to the ground within it. A throw is counted when any part of the fighter's body other than the soles of his feet touches the ground.

That initial breathtaking attack is intended to knock the opponent out of the circle. If that fails, the two men will grapple, each keeping as low as possible to remain stable and difficult to topple.

NODOWA

UCHI-GAKE

KOTE NAGE

UTCHARI

S T E A

Radar absorption
Triangles of radar-absorbe material, set into areas of aircraft such as wing leadi edges, trap radar energ within them, bouncin back and forth until i complet dissipate

A large crate on a trolley was hurried from the giant C-5 transport aircraft to an unobtrusive hangar. Inside, a select group of engineers immediately began dismantling the crate. Its cargo was an odd-looking aircraft. Those in the know called it 'Harvey', after the famous invisible white rabbit in the film. No-one else could see it . . . .

One o'clock in the morning, pitch black and deathly quiet outside. The hangar doors rolled slowly back. First

Heat and noise dissipation
Louvred fans on both engine intakes and exhausts dissipate heat and noise, both of which need to be kept to a minimum if the aircraft is to remain undetectable.

Heat shrouding
Vertical tail surfaces provide a large amount of radar cross-section in conventional aircraft. 'Stealth's' tail surfaces are small and turned inwards, shielding the engines from heat-seeking sensors.

Internal weapons
Any weapons are carried internally, as they contribute a large area to the overall cross-section if they are carried externally on pylons.

Heat absorption
The engine is surrounded by anechoic materials, which absorb the noise and heat.

out was a pick-up truck, with no lights save a barely visible strip light at the rear. It was followed by another, and they waited outside the hangar, engines purring. Last out was Harvey, an indistinct shape that melted into the dark background.

The only sound came from the rumble of the wheels as it glided over the taxiway. Within 30 seconds the convoy had reached the runway, the two trucks peeling off to the sides. Only the men in the trucks could hear the faint, low-pitched hum as Harvey's throttles were opened. By what little light was available on this

dark night, the pilot lifted off as quickly as a fighter and pulled the aircraft up in a steep climb. Within a minute of the hangar doors opening, Harvey was gone. Inaudible and invisible outside the confines of the base, the operation had been conducted with no radio, no lights, nothing to give away the secret departure.

Radio silence

The pilot worked the satellite navigation equipment to put him on course for the night's mission. He didn't have the normal navigation systems at his command: any radio or radar emissions from his aircraft could enable someone somewhere to detect him. These missions

were so secret that not even friendly military radar controllers could know about them.

The pilot dropped the aircraft down towards the deck as he approached the border, flipping on the infra-red and low-light-level television sets that would help him see where he was going, for this aircraft had no terrain-following radars. The plane skimmed across the border, the minefield strip separating East from West clearly visible. Far less visible to anyone on the ground was the aircraft, rushing silently towards its target, a chemical warfare works. The pilot monitored the instruments that detected any radar looking his way. Several were showing, but the remarkable shape of the aircraft and its radar-absorbing materials would not return enough of an image to alert the radar controllers. The specially-treated black paint returned no

LTH the invisible warplane

A squadron of super-secret warplanes are flying around the Nevada desert in the USA, unseen by the public – and non-existent according to officials. But they are invisible in another sense, for they are almost impossible to detect with radar and make almost no sound. They are known as 'stealth' aircraft, and they add a new, mysterious dimension to the secret air war.

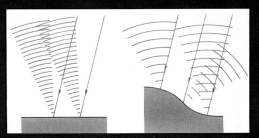

Radar diffusion
A flat surface reflects sharp beams of radar energy (much like a mirror reflects light). 'Stealth' aircraft feature blended curves that do not produce a sharp return, the curves flinging back the energy in many different directions.

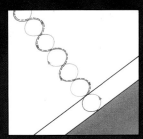

Phase-bending
Radar is a wave pattern, and some special materials on the aircraft's skin can reflect the radar wave back in two components that are exactly out of phase with each other, completely cancelling out the return.

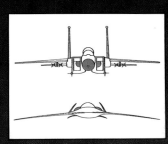

Radar cross-section
In a conventional aircraft such as the F-15 Eagle, the compressor blades on the engines and the radar dish in the nose give massive radar reflections. Other areas, such as the straight wing and fin leading edges and externally carried missiles, also reflect a large amount of energy. 'Stealth' has none of these features, and therefore has a vastly-decreased frontal radar 'cross-section'.

Satellite navigation
Navigation is handled by links with friendly satellites, enabling 'Stealth' to pinpoint its position at any time.

Passive sensors
'Stealth' has no radar, as this could be picked up by ground sensors. Instead it uses passive sensors such as infra-red and low-light level television. A laser radar may be employed for terrain avoidance.

glints, and matched the sky behind him. Carefully positioned shrouds masked any heat given off by the engines from infra-red sensors, while mufflers around the engines kept the noise to a faint whisper.

Tense moment

He was something of a veteran at this game, but every time he crossed that border, his gut strings tightened a notch. How long would this game last? How long before these 'stealth' techniques showed a vulnerable spot? He flew an unmarked aircraft over hostile territory, he wore no identifying patches or badges on his flight suit, and he carried no papers. The guys down there would sure make a stink if they knew.

To get the pictures and signals intelligence that he wanted, the pilot had to move in dangerously close to radar sites. Time to switch on Harvey's box of tricks. They called it DECM, or deception electronic countermeasures, which messed up the radar returns so much that a blip on the controller's screen appeared many miles away from where Harvey was. At times it returned the radar beams in phase with those coming in, completely blanking out the return. DECM kept him hidden by playing with the radars.

The navigation system had put him right over the chemical works, and the pre-programmed sensors automatically began blazing away at the target. Infra-red sensors took 'heat' pictures while electronic receivers lapped up the signals given off by running machines and telephone lines. Even the signal from truck ignition motors was measured. So efficient was the hardware that one run was enough, and the pilot set course for home.

The radar controllers below him reported another peaceful night. Sure, they'd had some odd returns, but no aircraft, just some birds flying about and a few atmospheric tricks that often happened on these hot nights. None of the tracks had looked like a hostile aircraft.

Back home

Harvey and its pilot ran before the wakening sky: they had to be put back to bed before the dawn revealed the secrets of this amazing aircraft to the world. The smooth landing felt reassuring to the pilot, as did the faint light of the pick-up truck waiting at the end of the runway. Together they raced for the small hangar. An hour later, the first cock-crow shattered the stillness of the countryside surrounding the chemical works. Its owners would never know that the world's most secret aircraft had flown over it, learning all the secrets of their capability for chemical warfare.

THE HAIRY HUMAN

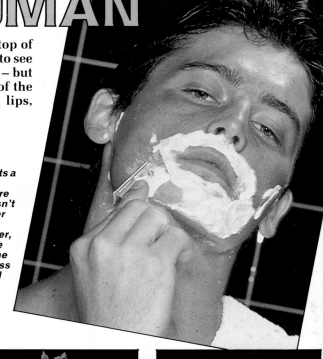

Hair covers every inch of your body from the thatch on the top of your head down to the fuzz on your toes. It may be difficult to see hair in some places – the inner side of your arms, for example – but it's certainly there. The only places without it are the palms of the hands, soles of the feet, backs of finger and toe tips, eyelids, lips, nipples and the navel.

There is a great difference between head hair, pubic hair and general body hair on any human being. There is also a great variety of hair types among individuals. Hairs vary in thickness, weight, colour, the length to which they will grow, and their shape in cross section. Straight hairs look round in cross section while curly hair tends to be oval.

Hair colour is due chiefly to the amount of melanin pigment in the hair cells. Whether your hair is naturally blond or dark, curly or straight, abundant or sparse, depends on the

The morning stubble gets a trim. It is often said that shaving encourages more beard to grow, but this isn't true; beards grow thicker and coarser with time because, as they get older, men produce more of the hormone that controls the distribution and thickness of hair. The world record beard belonged to a Norwegian and was 5.33 metres long.

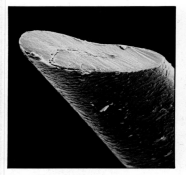

Traditional razor blades give the closest shaves, cleanly slicing through the hair shaft.

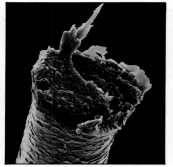

Scissors make more of a mess, rather like a blunt axe hacking through an old, tough tree.

Electric shavers devastate the beard, ripping through and leaving stumps like a forest after a tornado.

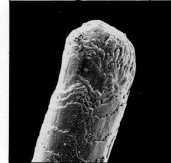

New, growing hairs have rounded tips of tough, but dead, cuticle. Despite their thinness they are very strong.

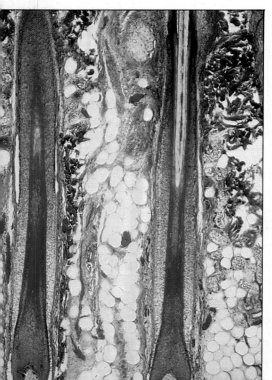

genes you inherited from your parents and grandparents. If you are male and your father is bald you are likely to go bald in the same way. If one of your parents is of African descent your hair is likely to be very curly or kinked.

The average head of hair contains 100,000 hairs, which remain in place for between two and five years. Up to 100 fall out and are replaced each day in a normally healthy person. Almost 50,000 hairs have to fall out before the hair is noticeably thinner. Eyelashes have a shorter life, of three to five months. At any one time roughly 10 per cent of hairs are dormant, while the others are growing.

Hairs under the microscope

Every hair grows from the base of a sheath, called a follicle, embedded in the middle layer of the skin. The base of the hair, the root, contains a few living cells but the shaft – the part we see above the skin – consists entirely of dead cells.

SOME COMMON MYTHS

● Hair can turn white overnight

Hair goes white very gradually with age, when hair cells lose their pigment. White hairs aren't black ones that have changed colour – they have to grow, and it takes months for a white hair to grow even three inches.

● Rubbing vitamin E into bald heads makes hair grow again

Little can be done to make the hair of baldies grow again, although there is one substance that has had some limited success. Desperate baldies can resort to surgery, where real or artificial hair is transplanted into the scalp, but the results are not always very convincing!

● Diseases can be diagnosed from hair

This belief is based on the fact that people with some diseases lack certain minerals, so their absence from the hair may indicate an illness. However, there is tremendous variation in the amount of minerals people have in their hair, and an experiment in which the same lock of healthy hair was sent to four laboratories produced four very different and inaccurate diagnoses.

The Venus Fly Trap

« The Venus Fly Trap is one of the few plants that can count! »

In the normal course of events, animals eat plants. But sometimes it's the other way round: when a plant can't get enough food from its roots in the ground, it makes sense to eat meat that arrives by air. The Venus Fly Trap lives in boggy areas where the stinking mud has few of the essential minerals needed by plants. But where there's stink, there's flies, and they are top of the menu for the Fly Trap.

"zzzzzzzzzzzzzzzzzz . . . Hmmm. Looks like a tasty corpse. Just what I fancy." Hard luck, fly; you're about to become a meal for an unusual green killer that uses tricks, mathematics, engineering and chemistry to earn a living where other plants fail.

The plant has evolved specialised pairs of leaves that are hinged in the middle and have long spikes at the edges. In the centre are blood-red patches that act as bait; flies are attracted to this colour because it usually means a meal for themselves and their maggots. The victim buzzes along to investigate. Sorry, fly – not a tasty corpse. Too late; it's triggered the trap and the blades have begun to close around it.

Those blood-red patches now start to ooze digestive juices that dissolve the captured fly. For a week or 10 days this external stomach will eat away at the victim. When every last piece of food has been digested the trap will open, leaving the empty carcass to blow away in the wind.

But plants don't have eyes, nerves or brains, so how does the Venus Fly Trap know when it's got a fly rather than a raindrop? It has a few sensitive hairs in the middle of the trap that seem to be able to count. A raindrop touches the hairs once; no response. A fly will touch them twice or more, and that triggers the trap.

The fly has triggered the hairs that identify him as dinner. Confused, the fly awaits the closing of the trap around him. The spikes act like prison bars. Soon digestive juices will ooze from the plant, dissolving the living fly.

Hairs a plant can count on

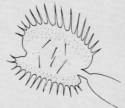

The sensitive hairs on the faces of the trap respond when they're stimulated more than once. Then they know they've trapped a fly rather than a raindrop.

A few days later and the feast is over. An empty fly skeleton awaits disposal by the wind.

THE FORGOTTEN
Jungle Survivor

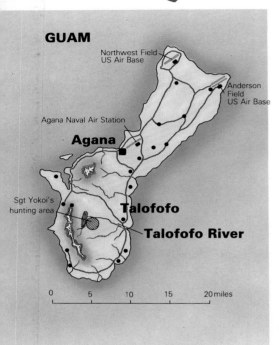

GUAM

Northwest Field
US Air Base

Anderson
Field
US Air Base

Agana Naval Air Station

Agana

Sgt Yokoi's
hunting area

Talofofo

Talofofo River

0   5   10   15   20 miles

Shoichoi Yokoi was taken to the town of Agana shortly after his capture. He is wearing the bark-fibre clothing he wove and made himself during his 30 years of solitude.

At dusk on 24 January 1972, two hunters were returning from the thick woods around the Talofofo River in the south of the Pacific island of Guam, when they came across a ragged scarecrow of a man. Covering him with their rifles, they marched him off and handed him over to the police. It was then that the amazing story of their strange captive began to emerge.

It had begun 28 years before, in the summer of 1944, when US forces were sweeping through the Pacific. Despite fanatical resistance, Guam finally fell on 10 August. Few prisoners were taken; more than 10,000 Japanese soldiers died fighting rather than suffer the shame of surrender.

One of those presumed dead was 28 year-old Shoichoi Yokoi, formerly a tailor's apprentice from Nagoya. He and two companions had in fact melted into the forests of southern Guam. Leaflets dropped by American aircraft told them that the war was over but, fearing they would be executed, they stayed in hiding.

They dug a cave, well-hidden amid the bamboo thickets, moving away from it – and then never very far –

only at night. For a few months they lived off supplies they had brought with them. Yokoi's companions then moved to another cave to fend for themselves.

Living off the land

Yokoi's first problem was clothing as, following orders, he had burned his uniform. Then he realised that the bamboo, palms and vines growing around him would yield fibres that could be twisted into thread. The former tailor wove these into cloth, which he stitched into garments.

Surprisingly, in a country that suffers 22.5 cm of rain annually, Yokoi's next – and constant – problem was finding drinking water. There were streams everywhere, but they harboured amoebic dysentery, cholera and typhoid as well as parasites, worms and leeches. Yokoi, however, found that certain vines and bamboos retained clean water in drinkable amounts. And in the wet season he used old bamboo stalks as a raintrap.

Plenty of fruit

Food was easier to find. Since arriving on the island he had eaten its nuts, breadfruit, mangoes and papaya. Coconuts provided both food and drink, while the shells made useful containers to supplement his army-issue rice boiler and kettle and the pan he had made from an old steel helmet.

Animal protein was harder to come by, although rats, snails and frogs were relatively common. Yokoi avoided the risk of being poisoned by the skin of certain frogs simply by skinning them before cooking. Occasionally he managed to trap a wild pig, but did once get food poisoning from one that he didn't cook for long enough.

Night fishing

With so many streams nearby, freshwater shrimps became a regular feature of Yokoi's diet. He also caught fish from the streams, using a net woven from tree bark or a trap of

woven bamboo. These he set at night, leaving them alone in the day and checking them next evening.

Seven years' solitary

As the years passed, Yokoi continued to visit his two compatriots occasionally. Then, sometime in the mid-1960s, he went to their cave and found them dead. Yokoi believed that they died of starvation.

For another seven years he continued his survival routine entirely alone, hiding by day, setting out at night to hunt and fish. He was on one of these expeditions when he was found by the two hunters. And so, after nearly 28 years, Shoichoi Yokoi came back to the world.

It was not an easy return. But there was one consolation – of a sort. As he had never officially left the army, the Japanese government decided he was entitled to his back pay. In 1944 he had received a wage of nine yen a month. For his 28-year war, Sergeant Shoichoi Yokoi was paid a grand total of 43,131 yen – which, at 1972 rates of exchange, came to about £54.

A day in the life of a fly

A fly's day starts soon after dawn. Flies are cold-blooded, but as soon as the sun starts to warm up the world, they begin to stir. The middle section of their bodies, the thorax, is full of powerful flight muscles, and on cold mornings the fly can flex those muscles very fast to warm itself up, just as we shiver when we are cold. If the fly is female, she starts to look for somewhere to lay eggs.

The mature female fly carries about 400 eggs at a time and produces them constantly, laying them in batches of up to 150 on almost any organic matter that is moist and decaying. Before the motor car replaced the horse, a favourite site was in the great stacks of horse manure in our towns and cities, but nowadays the major fly problem comes from the huge amounts of rubbish we produce and dump every day – although dog's excrement is highly favoured.

It is the heat produced by bacteria breaking down the rotting matter that allows the egg to develop and hatch as maggots in as little as eight hours. The maggots wriggle away from light, down into the warm decaying matter until they reach a layer at about 45°-50°C (113°-122°F). In really large compost heaps, the internal temperature can be as high as 70°C (158°F).

In this natural incubator, the maggots feed and grow in less than two

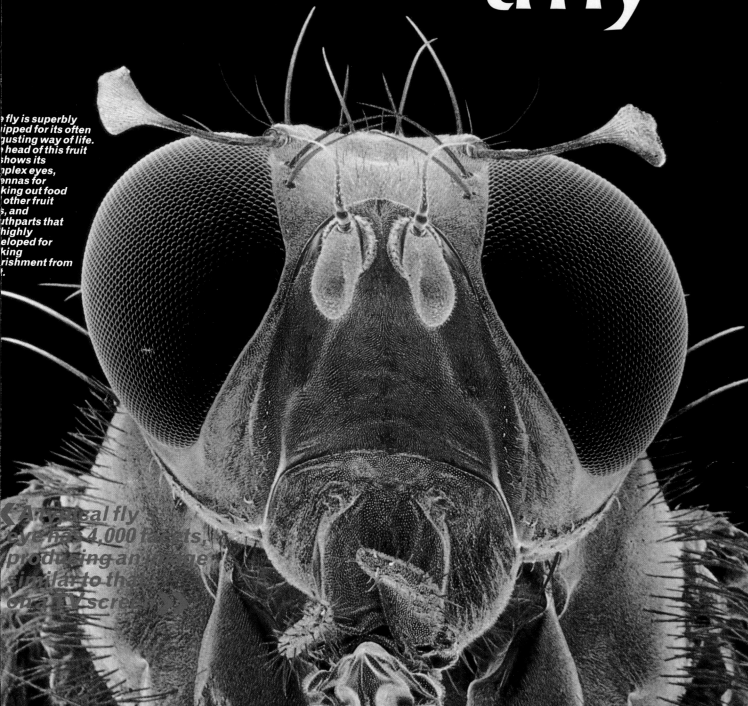

The fly is superbly equipped for its often disgusting way of life. The head of this fruit ... shows its complex eyes, antennas for ... king out food ... other fruit ..., and ... uthparts that ... highly ... eloped for ... king ... rishment from ...

‹ An typical fly eye has 4,000 facets, producing an image similar to that on a TV screen ›

This is the hairy tongue of a house fly. Flies are dependent on fluid food; if a meal isn't already liquid, then a quick vomit of digestive juices will soon make it so. There's no telling what the fly has last eaten but, whatever it was, there will be traces of it left behind on its new meal (maybe your dinner!).

Feeding time

Delicious smells fill the air when we are cooking, and the hungry flies find them irresistible. The fly's antennae act like a nose, sensing these smells, and soon the fly is at the kitchen window, looking for a way in.

A fly's major need is for sugar, so sweet foods are most popular. The fly lands on the food and 'tastes' it, using special taste buds, which it has on its feet! This saves time, as the fly can tell straight away whether the food is worth eating.

Once on suitable food, the fly extends its mouthparts, which are modified to handle liquid foods only – there are no piercing or chewing parts. In order to feed, the fly produces large amounts of saliva, which contains digestive chemicals. These break down the solid food into a liquid, which is then sucked up. Because of the design of its stomach, the fly vomits some of its last meal, too (possibly dog excrement), which all helps to digest the food. Flies also defecate regularly – about every five minutes – so the food is thoroughly contaminated.

About two days after emerging from the pupal case, the flies are sexually mature, and mate. The male usually finds the female by sight and the smell of the sexual chemicals, or pheromones, that she releases. The male springs onto the female's back, holds her with his legs, and strokes her head – if she is not ready to mate, she will kick him off. When she has been fertilised, the female flies off in search of somewhere to lay her eggs, and the whole cycle begins again.

weeks, moulting their skins three times before they prepare for the metamorphosis that will change them into adult flies.

The maggot's outer skin darkens and hardens, while all its internal structure is broken down, forming a nutrient 'soup'. Now small groups of cells which have previously been dormant start to divide and grow rapidly, using the soup around them as food. Each group is pre-programmed to make a certain part of the adult fly – some make the wings, others make the head, body, legs and so on. When the parts meet, they stop growing and fuse together. In four days, the fly is ready to break out of the protective skin (called the puparium). It pushes its way to the surface, expands its wings in the warm sunshine, and flies off to find some food.

These maggots are gorging themselves on raw meat. The ones on the right are only two hours old, and they've soon put on weight. Simple creatures, maggots; a mouth at the pointed end, along with two hooks with which to hang on, followed by a stomach. Because they liquefy their food by spewing enzymes, the group can feed more efficiently as a herd.

Below: Caught in motion, this housefly is just leaving its meal of brown bread. Despite appearances, flies have two pairs of 'wings', but the hind pair has been reduced down to tiny 'drumsticks' that vibrate to give the fly a sort of gyroscopic stability.

LANDING ON A CEILING

Not easy for other animals, but for flies it's a simple piece of aerobatics.

1. The fly flies parallel to the ceiling.

2. He puts up his first pair of legs.

3. The first pair of legs makes contact and sticks to the ceiling with its 'Velcro' hairs.

4. The fly stops flying and somersaults into position.

Above: Four thousand facets make up each eye of the typical fly, giving it a nearly spherical 360° vision. Each facet sends a signal to the brain, where a mosaic image of the external world is pieced together. The image is poorly focussed and in black and white, but it's good at detecting movement.

Fly feet each have a pair of pads that are covered in close-set, tiny hairs. These interact with the minute irregularities on ceilings, walls and windows, rather like a strip of Velcro, allowing the fly to grip.

A male fly sits astride a female while mating. He performs acts of stimulation to which his mate will hopefully respond, allowing him to have his way.

These puparia might look simple and boring, but inside each one an amazing transformation is taking place. The simple maggot has reduced itself down to a chemical soup. This soup reorganises itself in only four days to form a complete and perfect fly, one of the most complex insects on Earth. An adult fly can be seen emerging from its puparium case.

Right: After mating, female flies produce about 400 eggs in a steady production stream, laying them in batches of about 100. They need warmth to develop, so they are often laid in rotting vegetation or meat where decay produces heat.

The cellular telephone can do anything a conventional telephone can do, but with an added advantage – wires are replaced by radio waves, so the system can be used anywhere.

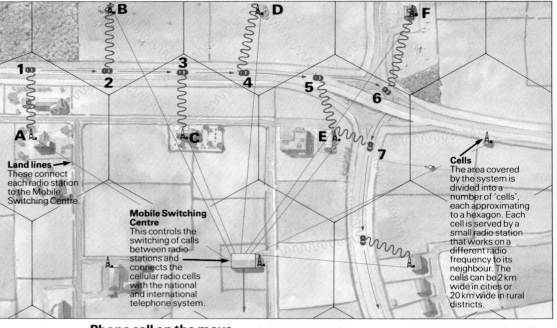

Land lines
These connect each radio station to the Mobile Switching Centre.

Mobile Switching Centre
This controls the switching of calls between radio stations and connects the cellular radio cells with the national and international telephone system.

Cells
The area covered by the system is divided into a number of 'cells', each approximating to a hexagon. Each cell is served by a small radio station that works on a different radio frequency to its neighbour. The cells can be 2 km wide in cities or 20 km wide in rural districts.

The country is divided up into a large number of radio 'cells', each of which is controlled by its own combined radio transmitter/receiver (transceiver), working on a number of frequencies. Each cell has to work on a different set of frequencies from the next one in order to avoid interference between cells. In this way, the network forms a repeating pattern of cells across the country with only a small number of frequencies.

When a cellphone user makes a call, he makes contact with the nearest cell transceiver and is given – by computer – a pair of frequencies on which he can speak and listen.

Phone call on the move
As the caller in the car moves along the motorway (**1**) he sets up a call with his nearest radio station, **A**. While he is making the call, adjacent stations (for example, **B**) are listening out and monitoring the strength and quality of his message and comparing it to the transmission they would be able to give.

Our moving caller has his call switched to station C immediately that station can handle the call more satisfactorily (position **3**). Switching

As he moves along the motorway to a position (**2**) where station B could handle the call better than station A, the call is switched via land line and computers to station B. Stations A and C now monitor the call.

continues as he moves through positions **4** and **5**, with stations **D** and **E** listening.

At position **6**, he should normally be transmitting via station E, but a concrete motorway flyover interferes with reception, so station **F** momentarily takes over. Once the vehicle is clear of the obstruction, transmission switches back to station E at position **7**.

Line of sight
Radio stations work on line of sight principles. If there is an obstruction such as a hill or large concrete structure, transmission is interrupted and an adjacent station takes over.

« Cellphones may soon be small enough to fit inside a wristwatch »

One of this pair is called a radio channel. The call is transmitted along this channel through the airwaves to the cell transceiver. This transfers the call to wires, which take it to an electronic mobile exchange (EMX), which in turn switches the call into the conventional telephone network. Once there, a call can go anywhere in the world. This means that any telephone in the world can be contacted from a cellphone.

When a call is made to a cellphone, either from a conventional phone or from another cellphone, the opposite happens: the call passes through the telephone network to the EMX. The system finds the nearest cell to the

cellphone and transmits the call to that cell. It is then allocated a radio channel and conversation can take place.

Down the highway

But what happens when a caller is moving along and he or she goes away from the cell transceiver? The system can handle that too. The strength of the signal he sends out is constantly checked. If the signal becomes weak, the call is switched over to another cell and given another channel. This process is known as the 'hands off', lasts less than one second and is virtually undetectable to the user.

Cell radio stations are often quite small, unobtrusive masts attached to existing pylons, blocks of flats or church spires. The larger Mobile Switching Centres are sometimes housed in office block basements or within small industrial units.

6 TO THE NORTH, HUGGING THE COAST OF GREENLAND, THE VICTOR III HAS RENDEZVOUSED WITH A MUCH MORE SINISTER SHAPE, A SOVIET DELTA IV ARMED WITH 16 SS-N-23 MISSILES, EACH WITH THREE OR MORE WARHEADS. IF IT COULD CREEP UP ON THE US COAST UNDETECTED, THOSE MISSILES COULD DEVASTATE THE WHOLE EASTERN SEABOARD WITH LESS THAN THREE MINUTES' WARNING. IT MUST BE STOPPED, NON-VIOLENTLY IF POSSIBLE.

7 SUPERB HAS PICKED UP THE APPROACHING SOVIET BOATS. SHE KNOWS THE TOLEDO IS AROUND, BUT SHE CAN'T HEAR THE QUIET, DRIFTING AMERICAN BOAT.

8 TO LET THE SOVIETS KNOW THEY HAVE BEEN DETECTED, SUPERB USES ITS ACTIVE SONAR, SENDING OUT PULSES OF SOUND ENERGY TO FIX THE ENEMY'S POSITION.

≪ *111 days* is the longest-ever submerged operational patrol by a submarine, carried out by HMS *Warspite* in the South Atlantic after the Falklands conflict of 1982 ≫

Sail structure
The sail (previously known as the fin, or conning tower) contains access to the control station, from where the boat can be conned (guided, or steered) while on the surface. It also houses the numerous sensors for operations on the surface or at periscope depth. These include an ESM (Electronic Support Measure) mast, used to detect enemy radar and communications; a radar mast with a surface search radar scanner; and a search periscope aft of a much more slender attack perisope, both of which can be used in conjunction with a number of electro-optical instruments, including infra-red, low-light TV, image intensifiers, thermal imagers and laser-ranging systems. The rearmost masts are 'snorts' through which surface air can be drawn if the boat has to run submerged on diesel power in an emergency.

Aft escape tower
There is an escape tower aft as well as forward, designed for the evacuation of the engineering spaces or for the rescue of crewmen in the machinery spaces, which are effectively cut off from the rest of the boat by the reactor.

Turbines
Steam, heated by the reactor, is piped at high pressure into the engine room, where it is used to power a pair of geared steam turbines. These are basically large fan blades turned by the steam at very high speed, and connected to the propeller through a gearbox, clutch and main drive shaft.

...rol centre
...ontrol centre is where the ...-day running of the boat ...s. The diving officer ...ols the ballast and trim ... and the planesmen ...euvre the boat on the ...in's instructions. On the ...ation console against the ...khead, the submarine's ...on is plotted. A highly ...sticated inertial navigation ...m is used while ...erged, but on the surface ...tes can fix the position to ... a few metres. The officer ... watch also works here, ... the various surface ...rs fitted to the masts ... up through the fin.

...ew quarters
...e crew quarters of a nuclear ...t are much more ...nfortable than those of a ...sel/electric boat, but the ...g patrols, lasting up to three ...nths, are still a strain. The ...w live below the control and ...ck centres, and have ...ilities such as galleys, ...shing machines, freezers, ...d so on. Officers are housed ...the control deck level.

Reactor space
The nuclear reactor, by far the heaviest piece of machinery aboard an atomic-powered submarine, is usually located at the centre of gravity. The reactor generates heat, and this is used to make steam to drive the turbines. Access through the reactor area is by heavily-shielded tunnels to the aft of the boat, which is devoted to engineering.

Auxiliary machinery
The space immediately aft of the reactor is occupied by the auxiliary machinery rooms, and on the lower deck a turbo-generator provides electric power. The diesel generator room on the level above is designed to drive the boat in the event of reactor failure. The upper deck, called the manoeuvring room, contains the engine and reactor controls and monitors; the tunnel through the reactor space leads from here to the control room.

nic gear to jam enemy radar, detect hostile radar signals, and so on.

The inner sanctums

The pressure hull is divided into two separately pressurised compartments by a bulkhead, so that, if one section of the sub is hit, the crew can move into the undamaged section and through escape towers leading between the hulls to the sea. Much below 30 metres this is extremely dangerous, and an emergency rescue vessel equipped with mini-submarines and diving bells is called in to give specialised help.

An SSN is divided into three main compartments along its length. Towards the stern are the steam generator, power turbine and turbo generator that turn the sub's massive multi-bladed propeller and produce the

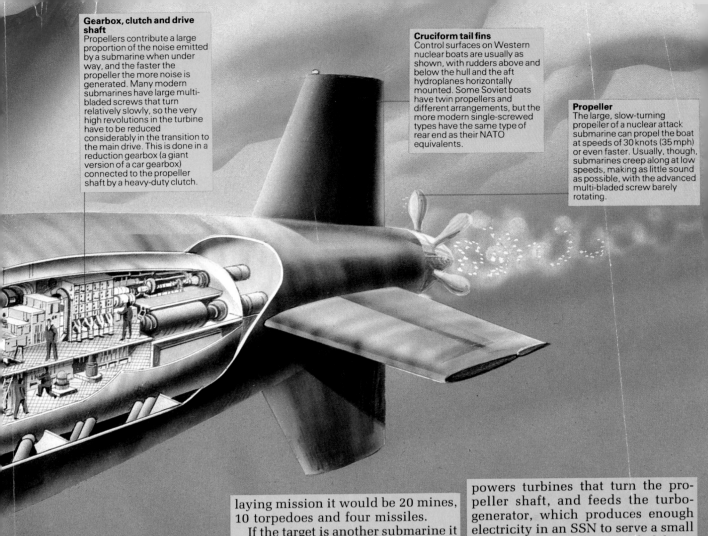

Gearbox, clutch and drive shaft

Propellers contribute a large proportion of the noise emitted by a submarine when under way, and the faster the propeller the more noise is generated. Many modern submarines have large multi-bladed screws that turn relatively slowly, so the very high revolutions in the turbine have to be reduced considerably in the transition to the main drive. This is done in a reduction gearbox (a giant version of a car gearbox) connected to the propeller shaft by a heavy-duty clutch.

Cruciform tail fins

Control surfaces on Western nuclear boats are usually as shown, with rudders above and below the hull and the aft hydroplanes horizontally mounted. Some Soviet boats have twin propellers and different arrangements, but the more modern single-screwed types have the same type of rear end as their NATO equivalents.

Propeller

The large, slow-turning propeller of a nuclear attack submarine can propel the boat at speeds of 30 knots (35 mph) or even faster. Usually, though, submarines creep along at low speeds, making as little sound as possible, with the advanced multi-bladed screw barely rotating.

ectricity that keeps the boat's systems operational. Heat to drive this machinery comes from a pressurised water-cooled reactor set roughly midships and aft of the fin. A heavily shielded tunnel runs through the reactor section to join the fore and aft working spaces. In the forward compartment are living quarters for the 100 or so crew, operations rooms, sensor compartments and the boat's array of armaments.

Choose your weapon

In a British SSN the bow is taken up by six 533 mm-diameter torpedo tubes and the long-range sonars. These are both 'passive' – that is, highly sensitive sound detectors – and 'active', which send out a carefully aimed blip of noise to bounce off a target and find its exact range and speed. Behind these is the torpedo room where underwater-launched antiship missiles and, occasionally, sea mines are held as well as torpedoes. A typical mix of armaments for a patrolling SSN would be 18 torpedoes and eight missiles; on a mine-laying mission it would be 20 mines, 10 torpedoes and four missiles.

If the target is another submarine it will be attacked with a wire-guided torpedo. Once launched, this is guided close to the target by commands sent down a wire connecting it to the sub. Once the target is within range of the torpedo's own sonars, these home it in for the kill.

If the target is a surface ship, there are several options. The sub can fire a sonar-equipped torpedo that homes in on the noises the target sends out; a direct-fire torpedo with no homing system and whose accuracy depends on the skill of the men firing it, since it has to hit the target or pass under it before it explodes; or an underwater-launched missile, which can be fired at a great distance from the target. It climbs above the surface and uses its own radar to lock on to the target.

Heart of the hunter

The key to an SSN's effectiveness is its nuclear reactor. Fuelled by cans of radioactive uranium-235 in its core, the reactor heats pressurised water to above its normal boiling point. This is then pumped into the coiled pipes of a large steam generator to heat up water surrounding the coil to boiling point. The steam pressure in turn powers turbines that turn the propeller shaft, and feeds the turbo-generator, which produces enough electricity in an SSN to serve a small town. The steam is then cooled down again, and the resulting water is recycled into the steam generator.

The reactor produces enough power to let the boat stay submerged for months on end if need be – and travel at top speed all the time. Air is kept breathable by chemically extracting poisons such as carbon monoxide and carbon dioxide, while fresh oxygen is generated by electrically extracting it from distilled seawater. Excess hydrogen, which is highly flammable, is vented into the sea. The only real limits on an SSN's patrol capacity are thus the amount of food it can carry and the health and morale of its crew.

Deadly and elusive, the hunter-killer submarine is now the first line of defence for both East and West, and is being constantly developed and refined. New features in the pipeline include water-jet propulsion, hulls that swallow up active sonar signals, further automation in weaponry and super-sensitive detector systems, and laser communication that will allow submarine commanders to talk to base via satellite even while fully submerged.

SUB CHASE

Deep beneath the storm-tossed waters of the North Atlantic, a potentially deadly game of hide and seek is being played. Submariners of the UK and the USA on one side and the USSR on the other are stalking each other in those sunless depths. Nowhere is the line between war and peace so finely drawn, and nowhere are the fighting men so ready for action. If war started, it could easily be like this . . .

1 TOLEDO, AN ATTACK SUBMARINE OF THE US NAVY, HEADS OUT FROM THE VIRGINIA COAST BOUND FOR HER NORWEGIAN SEA PATROL AREA. ONCE SUBMERGED, SHE WILL NOT REAPPEAR FOR 60 DAYS OR MORE.

2 A SHORT TIME LATER, AND A LONG WAY FURTHER NORTH, A SOVIET 'VICTOR III' ATTACK SUBMARINE LEAVES THE HIGH-SECURITY NAVAL BASE AT POLYARNY, ON THE KOLA PENINSULA IN THE SOVIET UNION.

5 ABOARD THE TOLEDO, PASSIVE SONAR OPERATORS LISTEN TO THE INNUMERABLE SOUND SIGNALS IN THE WATER, FILTERING OUT NATURAL NOISES AND ANALYSING THE REMAINING MAN-MADE SOUNDS.

3 HMS SUPERB, AN IMPROVED 'S' CLASS SUBMARINE OF THE ROYAL NAVY, SUBMERGES AS SHE LEAVES DEVONPORT ON HER WAY TO THE VITAL WATERS OF THE GREENLAND-ICELAND-UK (GIUK) GAP. THESE NARROW WATERS FORM A 'CHOKE POINT' THROUGH WHICH SOVIET VESSELS HAVE TO RUN THE GAUNTLET OF NATO DEFENCES IN ORDER TO REACH THE TRANSATLANTIC SHIPPING LANES.

SOSUS (NATO'S SEA-BED LISTENING SYSTEM) HAS DETECTED ONE OR MORE SOVIET SUBMARINES PASSING THE NORTH CAPE OF NORWAY. A PRE-ARRANGED CODE SIGNAL BRINGS SUPERB TO PERISCOPE DEPTH, WHERE SHE CAN RAISE HER VHF RADIO AERIALS AND BE CONTACTED DIRECTLY VIA COMMUNICATIONS MILITARY SATELLITE.

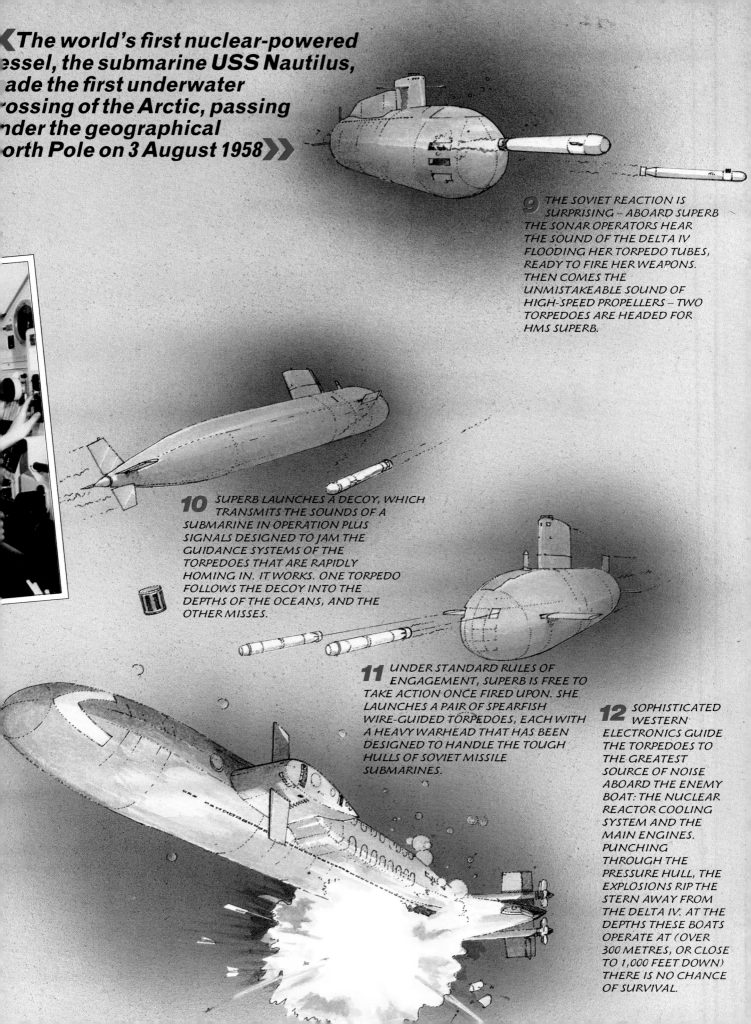

The world's first nuclear-powered vessel, the submarine USS Nautilus, made the first underwater crossing of the Arctic, passing under the geographical North Pole on 3 August 1958 ❯❯

9 THE SOVIET REACTION IS SURPRISING – ABOARD SUPERB THE SONAR OPERATORS HEAR THE SOUND OF THE DELTA IV FLOODING HER TORPEDO TUBES, READY TO FIRE HER WEAPONS. THEN COMES THE UNMISTAKEABLE SOUND OF HIGH-SPEED PROPELLERS – TWO TORPEDOES ARE HEADED FOR HMS SUPERB.

10 SUPERB LAUNCHES A DECOY, WHICH TRANSMITS THE SOUNDS OF A SUBMARINE IN OPERATION PLUS SIGNALS DESIGNED TO JAM THE GUIDANCE SYSTEMS OF THE TORPEDOES THAT ARE RAPIDLY HOMING IN. IT WORKS. ONE TORPEDO FOLLOWS THE DECOY INTO THE DEPTHS OF THE OCEANS, AND THE OTHER MISSES.

11 UNDER STANDARD RULES OF ENGAGEMENT, SUPERB IS FREE TO TAKE ACTION ONCE FIRED UPON. SHE LAUNCHES A PAIR OF SPEARFISH WIRE-GUIDED TORPEDOES, EACH WITH A HEAVY WARHEAD THAT HAS BEEN DESIGNED TO HANDLE THE TOUGH HULLS OF SOVIET MISSILE SUBMARINES.

12 SOPHISTICATED WESTERN ELECTRONICS GUIDE THE TORPEDOES TO THE GREATEST SOURCE OF NOISE ABOARD THE ENEMY BOAT: THE NUCLEAR REACTOR COOLING SYSTEM AND THE MAIN ENGINES. PUNCHING THROUGH THE PRESSURE HULL, THE EXPLOSIONS RIP THE STERN AWAY FROM THE DELTA IV. AT THE DEPTHS THESE BOATS OPERATE AT (OVER 300 METRES, OR CLOSE TO 1,000 FEET DOWN) THERE IS NO CHANCE OF SURVIVAL.

LIGHTNING STRIKE

Bolts of lightning engulf New York City during an electric storm. A single bolt can discharge up to 100 million volts in one hundredth of a second, arcing through the sky in a line only 5 mm wide at speeds approaching 5.5 million km per hour. This channel of energy isn't the lightning stroke itself but its 'step leader'; the lightning you see is a streamer of positive electricity rising from the ground to the cloud, back along the leader. Its temperature is about 28,000°C, and it travels at an incredible 126,000 km per second – nearly half the speed of light.

Somewhere on the Earth's surface lightning is always flashing – up to 100 times a second by some estimates, a non-stop power discharge of around four billion kilowatts. Lightning is the greatest of the weather killers, striking down far more victims than tornadoes or blizzards: some 200 people a year are killed by lightning in the USA alone. Death and injury come mainly from 'side-flashes' given off when high objects such as trees are struck, and from the immense voltages in the earth in the vicinity of a ground strike. Over half of those who die are engaged in a recreational activity, with golf at the top of the list – golfers are particularly at risk at the top of their backswing when using a metal club!

Killer strikes

The rapid compression and decompression of air near a flash can knock people flat on their backs and even rip off clothing. Death occurs when the heart, or certain parts of the brain, are on the route of the electrical discharge. Cows are more likely to die than humans, as the current passes through and between their wide-spaced legs, often travelling via the heart. Britain suffers less lightning than the USA, with an average of six strikes per square mile per annum. In April 1979, at Caerleon in

Wales, 11 soccer players leaving a pitch because of a storm were struck by lightning, but only one was seriously injured.

The sheer power of lightning is nearly incomprehensible. Prior to the flash, 100 million watts build up between the opposing electrical poles of the thundercloud. On release, the return stroke blasts skywards at a phenomenal 422 million feet per second, creating a temperature of up to 30,000°C in a conductive channel that may be pencil-thin but five or more miles long. Such power reduces wooden masts to matchwood, melts holes in church bells and welds chains into iron bars. It can

On 10 August 1975 lightning fused the metal artificial leg of a cricket

« Ball lightning consists of spheres of intense light that float through the air near thunderstorm disturbance. In 1938 a ball of lightning traversed the length of an airliner flying through stormclouds over France. The ball singed the pilot's eyebrows after coming through an open window, rolled the length of the passenger cabin, and then exploded harmlessly »

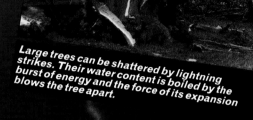

Large trees can be shattered by lightning strikes. Their water content is boiled by the burst of energy and the force of its expansion blows the tree apart.

cook root vegetables in the earth, shatter damp masonry, and cause trees to explode like bombs. Sometimes, when it strikes quartz sand, it produced 'fulgurites' – thin tubes of spontaneously-created glass up to five feet long, embedded vertically in the sand.

Rumbling thunder

The explosive discharge of energy and violent expansion of air of a lightning streak produces the detonation we recognise as thunder. It is actually a sudden bang, but as the sound waves are bounced off other clouds and obstructions it becomes a deep rumble. Sound travels at about 1 km

Poor Julius Saxton of White Plains, New York, saw his house blown apart in less than a second during a storm in 1936. Lightning delivers a huge amount of energy, and unless it can be immediately discharged to earth it is translated into explosive force.

ear Berwick-on-Tweed into a solid bar »

77

LIGHTNING STRIKE

Nobody knows for certain, but scientists have a theory of how lightning occurs.

1 As the build-up to a thunderstorm, an updraught sweeps warm air upwards from the ground. The air cools as it meets the colder atmosphere and condenses into droplets of water, forming clouds.

2 As the updraught continues, travelling at up to 60 miles per hour and reaching 15,000 metres in height, the clouds become taller. Water droplets towards the top are supercooled to form ice crystals; as they fall they create a downdraught, and sometimes come down as hailstones if they become too heavy for the surrounding air currents to hold up.

3 Static electricity is created within the cloud as the various drops of water are agitated against each other by the moving air currents – positive electricity towards the top of the cloud, inside the ice crystals, and negative down towards the base, as the crystals jostle against the warmer drops.

4 The negative charge attracts positive charges in the ground below, and – nobody knows why – a small pocket of positive charge nestles at the bottom of the cloud. The two are unable to meet because of the intervening cushion of air, and soon the cloud is a seething mass of electricity.

5 The critical point is reached when the electrical forces in the cloud and those in the ground are strong enough to break through the insulating barrier of air and form a violent connection. A 'wire' of negative electrons, up to 100 million volts, shoots to the ground in less than a thousandth of a second, zig-zagging on its downward path. This is the 'step leader'.

6 The bolt you actually see is the second stroke, returning along the same path as the step leader and glowing visibly. 'Sheet' lightning is a reflection – of single lightning bolts against clouds and the atmosphere. Often there are multiple strikes along the same path, so you see three or four flashes in rapid succession.

7 The lightning bolt makes the sound we know as thunder, when a shock wave surges out from the hot, and thus expanded, air around the bolt. It is actually a single sound, but the rumbling we usually hear is caused by the sound echoing off other clouds and the ground.

8 The lightning bolt will take the shortest route to Earth, and so the highest point in that area will be the line that gets struck – a lone tree in a field, a tall building – or a human being!

9 It is easy to calculate how far you are from a thunderstorm. Sound travels at about 1 km every three seconds, so if you count the number of seconds between seeing the lightning flash and hearing the thunder and divide by three, this will give you the distance in kilometres.

in three seconds, so the distance of the storm can be judged by counting the seconds between flash and sound and dividing by three.

Lightning will always discharge to earth through the highest available point, such as a tall building or solitary tree, making it unwise to shelter under the latter during a storm. Even if you are not hit by a direct strike, the power of the streak can instantly vapourise water inside the tree, causing it to explode dramatically.

Strangely enough, a direct strike on a human need not always be fatal or even dangerous. Our high water content makes us good conductors of electricity, and the current will flow through us to the ground; providing that it does not pass through the heart or brain, there is a good chance of survival.

CAUGHT IN A STORM

If you're indoors:
Stay there. If you have laundry hanging on a metal clothesline, leave it there.

Stay away from open doors and windows, fireplaces, radiators, stoves, metal pipes.

Don't use plug-in electrical equipment of any kind – and don't pull the plugs out during the storm. This includes the telephone!

If you're outside:
Keep away from – and certainly don't work on – fences, phone or power lines, pylons, metal piping, or steel scaffolding.

Don't use metal fishing rods, golf clubs or other metal sports equipment. Golfers wearing shoes with metal cleats make excellent lightning targets.

If you're in a tractor, get out of it, especially if it's towing a plough or other metal appliance in contact with the ground.

Try to find a building to shelter in. If it's metal, stay away from the walls. Caves are dangerous; humans are better lightning conductors than stone walls!

Don't shelter under isolated trees; an open field is safer, but crouch down. Stay at least twice as far from the nearest tree as its height.

If you're travelling:
Cars and aircraft, bikes and motor bikes are usually safe in a storm, but don't touch the metal exterior or frame.

If you're in a small boat or out swimming, get on to dry land as soon as possible.

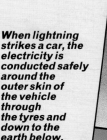

When lightning strikes a car, the electricity is conducted safely around the outer skin of the vehicle through the tyres and down to the earth below.

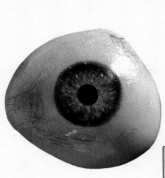

MEET THE BIONIC MAN

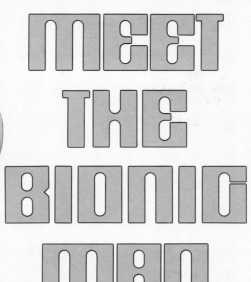

From artificial blood-cells that outperform the real thing to synthetic thigh bones that lengthen as the body grows, bionic replacement is flourishing as never before. The bioengineers now have the materials and achievements of space and microprocessor technology to help them. The idea of advanced bionics is familiar to millions of viewers through TV and film fantasies such as the *Six Million Dollar Man* and *The Terminator*, and man-made body parts and systems are a reality for the men, women and children who would be crippled or dead but for bionic replacements. The bioengineers and surgeons who spearhead research are not attempting to create the superhuman abilities of the screen heroes and villains; instead they are concerned to repair and service the frail human body by recreating and applying hi-tech versions of its own complex and efficient systems.

Electric impulses

The body does not have to be taught how to use the latest bionic limbs – it already knows. The mental or reflex decision to perform a physical action causes electric impulses known as electromyographic signals to stimulate the necessary muscle action. Bionic engineers utilize these signals, which still exist although the limb, or part of a limb, may have been lost. In the 'thinking arm' developed in Sweden and the USA, electrodes attached to the arm stump and to the patient's shoulder, chest and back on the side of the damaged limb pick up the electric signals and transmit them to a minicomputer situated in the artificial arm.

Wear and tear

Accidents aside, certain parts of the human body wear out faster than others, and their replacement is a major concern of bioengineering research. Hip and knee joints are a prime example; most of the 190,000 joints replaced annually by US surgeons are hips. Replacements are constructed of strong, lightweight metals like chrome-cobalt or titanium alloy, moving against tough plastics like polyethylene. The device is cemented to the existing bone.

This has been fine for the less active, elderly patients who make up the majority of replacement candidates, but a younger person needing a replacement joint will make greater physical demands, and in such cases surgeons are beginning to use bionic replacements with porous surfaces into which the bone will grow.

« **A Canadian amputee was fitted with a myoelectric arm specially modified so that he could play the saxophone!** »

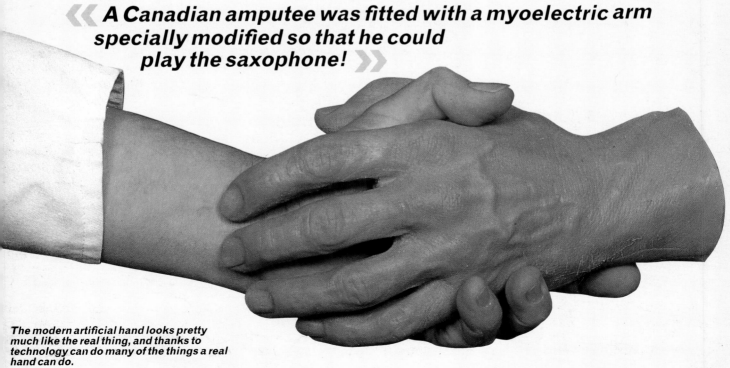

The modern artificial hand looks pretty much like the real thing, and thanks to technology can do many of the things a real hand can do.

Replacement hearts

The human heart has long been the subject of replacement research and practice, and these days bionic heart components are made on the same sort of vacuum mould as is used for producing plastic cups. The most successful replacement heart in recent years, the Jarvic-7, weighs almost two pounds, but a prototype heart has now been produced weighing only a pound, which is about the same weight as a real heart. This development, by Pennsylvania State University, responds to varying blood demands, which occur with different levels of activity and with strong emotions like fear or rage. The problem with artificial hearts and other internal organs is that they need a power-pack, and, so far, external 'invasive' systems have been used, working on battery-stored electricity or on compressed air, both of which require through-skin connections.

Eye problems

While the eyes of TV's 'Bionic Man' could be used like high-powered telescopes and had infra-red night vision, most blind people would be satisfied with a replacement that gave them vision of any sort. This is a problem area for bio-engineers. The eye is really an extension of the brain, and even if a workable device could be manufactured it would have to be somehow attached to the staggering two million nerve fibres that connect the eye to the brain. Knowing that electrical stimulation can produce 'visible' light spots, called phosphenes, in some blind people, researchers have used electronics to 'draw' pictures in a blind man's mind. A 5-cm square piece of Teflon patterned with 64 electrodes was implanted between the lobes of the subject's brain, and connected to a computer via wiring through the skull. As a TV camera picked up images, they were transmitted via the computer to the electrodes, and the subject 'saw' well enough to read sentences.

Hearing aid

Electric impulses are also the principle behind the cochlea implant, a bionic device designed to help the deaf hear again. Sound is translated into electricity, which activates stimulation of the auditory nerve. Tonal quality is missing, but the new ability to hear sound can radically alter the lives of those imprisoned in the silence of old age.

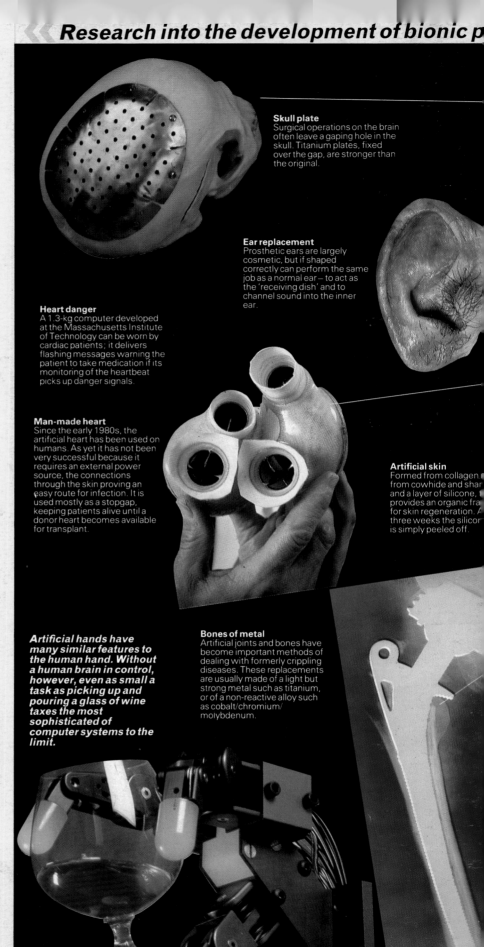

Skull plate
Surgical operations on the brain often leave a gaping hole in the skull. Titanium plates, fixed over the gap, are stronger than the original.

Ear replacement
Prosthetic ears are largely cosmetic, but if shaped correctly can perform the same job as a normal ear – to act as the 'receiving dish' and to channel sound into the inner ear.

Heart danger
A 1.3-kg computer developed at the Massachusetts Institute of Technology can be worn by cardiac patients; it delivers flashing messages warning the patient to take medication if its monitoring of the heartbeat picks up danger signals.

Man-made heart
Since the early 1980s, the artificial heart has been used on humans. As yet it has not been very successful because it requires an external power source, the connections through the skin proving an easy route for infection. It is used mostly as a stopgap, keeping patients alive until a donor heart becomes available for transplant.

Artificial skin
Formed from collagen from cowhide and shar and a layer of silicone, provides an organic fra for skin regeneration. A three weeks the silicon is simply peeled off.

Artificial hands have many similar features to the human hand. Without a human brain in control, however, even as small a task as picking up and pouring a glass of wine taxes the most sophisticated of computer systems to the limit.

Bones of metal
Artificial joints and bones have become important methods of dealing with formerly crippling diseases. These replacements are usually made of a light but strong metal such as titanium, or of a non-reactive alloy such as cobalt/chromium/molybdenum.

‌UMAN ACCESSORY MANUAL

A false smile
False teeth were amongst the first prosthetic devices developed by mankind. They are also the commonest, especially in the sweet-consuming societies of the 1980s.

Silicone implant
Cosmetic surgery, often undertaken for vanity rather than for any medical reason, will sometimes use artificial material. This is a silicone implant, used to enlarge female breasts.

A hand for tomorrow
Computer research has revolutionised the development of artificial hands. Micro-electronic switches, controlled by the muscle twitches in the arm or shoulder, for instance, feed commands through a micro-computer to the powered hands.

Bionic knees
Surgeons are now combining radiographic techniques with computer-generated 3-D graphics to design 'ideal' bionic knees, using a data-base of hundreds of normal knees to predict what the natural growth of an arthritic knee would have looked like.

New legs for old
Legs tends not to be as sophisticated as arms, as they don't have to do such sensitive jobs. Nevertheless, modern legs are the subject of much research. They are being made of lightweight materials such as Kevlar, and research into the most efficient system of joints and joint lubrication is going ahead fast.

❮❮ *Some patients have had as much as 80 per cent of their major arteries replaced with Dacron and Teflon substitutes* ❯❯

VELCRO

Space Age Fastener

Sometimes the simplest of ideas are the most revolutionary, and often they come from taking a close look at nature. Velcro, the original hook and loop fastener that has made buttons, zips and a host of other mechanical fasteners redundant for many applications, started life in the 1950s when its Swiss inventor, out hunting, had to remove the tenacious, multi-hooked burrs of the burdock plant from the fur of his dog and the woollen fibres of his own clothing.

The resulting invention has travelled into the depths of space on the multi-million dollar spacesuits of NASA astronauts, regularly circles the world in Concorde and other jetliners, and has helped protect mountaineers from the winds and deluges of high terrain from Everest downwards. On the mundane level, Velcro tapes and similar hook and loop fasteners are found in most households, cars, factories and hospitals.

« One type of Velcro has a shear strength 10 times greater than the force needed to peel it apart »

The nerve-grating ripping sound of Velcro being peeled apart occurs as the thousands of nylon hooks of one strip pull free from the equally numerous loops of the opposing tape.

This electron micrograph picture, magnified 14 times, shows the numerous hooks and loops of Velcro tape. The bond is easily broken by the vertical pull of a peeling action, but its enormous lateral shear strength and ease of operation makes it ideal for use in the most demanding circumstances – including that of space travel.

A close-up reveals the hooks and loops. Velcro is manufactured in two separate parts; one tape has a hooked surface, made by weaving loops and then cutting them, and the other tape is a series of loops.

She just stepped out........"

It's an all too common scene. A white-faced car driver, shaking with shock, is seated in the back of a police car with his head in his hands. His car is slewed sideways across the road at the end of an uninterrupted line of skidmarks. A little further on, ambulancemen lift a stretcher into their waiting vehicle. There is blood on the road.

"Honestly, I didn't see her until she stepped out in front of me," the driver keeps repeating. "I wasn't going fast, but she didn't give me a chance!"

The police have heard it all before. They know these scenes only too

Laden with shopping, the woman steps out into the road – and into the path of an oncoming car. The driver is going too fast; he takes his eyes off the crossing, and the result is fatal.

well, and they know that the driver is telling the truth as he sees it. They also know that there will be almost as many versions of the truth as there were witnesses to the accident. So how can the *real* sequence of events be worked out?

There *is* a method, one which makes use of the laws of physics rather than the uncertain recollections and vivid imagination of ordinary human beings. Through measurement and calculation, police accident investigation specialists can work out the exact sequence of events leading to a tragedy.

THE METHOD
The method the police use is that of reconstructing the accident backwards from the resting place of the car. By computing the braking distance of the car and the reaction distance of the driver, the police can work out whether or not the driver was driving dangerously.

Braking distance
All vehicles, from Mini to double-decker bus, take the same distance to stop on dry roads. The length of the skid mark tells exactly how fast the vehicle was travelling when the brakes were slammed on. In this case, the skid is 18 metres long, which is the distance needed to stop from a speed of 34 mph.

Reaction distance
Most people take 0.7 seconds to react to a situation on the road, lift their foot off the accelerator and stamp down hard on the brake. In that 0.7 seconds, a car travelling at 34 mph will travel just over 10 metres.

Braking distance 18 metres

Point of impact

Reaction distance 10 metres

Driver sees pedestrian

THE CONCLUSION
The sequence now becomes clear. The driver took his eyes off the road for perhaps 1½ seconds, just as the woman stepped on to the crossing. When the driver looked back again, it would have indeed seemed that the victim "just appeared" in front of his car. In fact, his momentary lapse of attention was the major cause of the tragedy.

The high-speed lady
If, as the driver swears, the lady "just stepped out in front of me", then she must have done so quite fast. To cover the 3 metres from the pavement to the impact in the 0.7 seconds after the driver saw her, she must have been running at about 12 mph, which is a pretty good speed even for a fit jogger on his daily run!

Point of impact
The skid mark shows where the car went over the crossing, and in combination with the dent in the car tells the accident investigation team the distance of the victim from the kerb (in this case, about 3 metres).

Driver sees pedestrian
By adding together the braking distance and the reaction distance, accident investigators can identify the point at which the driver *saw* the pedestrian.

85

INTO THE FUTURE
Lift-off to Saturn

In the year 2050, a handful of intrepid astronauts set out in the biggest, most advanced interplanetary spacecraft of their era on an astonishing voyage – a six-year, 1,400 million-km journey to deliver a prefabricated research station to the rings of Saturn.

Part One: The First Stage – from Shuttle to Takeoff

Fitting out
A complete Lunar-class cargo ship runs pre-launch fitting-out checks. She is destined for the triangular routes between Earth orbit, the Lunar stations and the giant space colonies between Earth and Moon. Her capacious cargo bay carries manufactured goods from the orbital industrial complex, raw materials from the Moon, and foodstuffs from the 10-km long cylinders that are the space settlements in orbit more than 30,000 km above the industrial belt. Similar vessels, with larger drive sections, make the run to Mars; automatic unmanned models collect from the rich mines of the asteroid belt.

Construction
The basic skeleton of an asteroid m ship begins to take shape. Statione permanently in space and with no r fight planetary gravities, these vess be of much lighter construction tha surface-bound craft like the Shuttle version of the asteroid ship is used Mars run, the only difference being addition of a landing module to get from the surface of the Red Planet.

The shipyard
As the Earth shuttle approaches, a number of different vessels can be seen at various stages of completion in the Interplanetary Engineering and Laboratory Corporation's orbital construction dock.

Maintenance
An Earth shuttle undergoes routine maintenance in one of IEL's repair ba Cheap solar energy, skilled engineers high degree of automation make such operation much cheaper up in orbit th would be 'down there'.

Ship finishing
The framework of an Outer Planet Explorer-class vessel is complete. Now the outer skin of the spin module is being fitted along with one of the ship's huge solar power panels.

Five of the six men and women in the standby room seemed tense as they waited for the signal to board the orbital shuttle whose looming shape could be seen through the observation port, that would take them to the giant spacecraft already in Earth orbit in which they would make their vast journey. All were experienced spacecrew, but they were still only the second team to face the 1400 million-km flight to Saturn. The journey was still fraught with uncertainties despite the lessons learned from the first manned expedition.

At least those first pioneers would be there to greet them. Meanwhile, as each of the five were well aware, they would have only themselves to rely on for the six years of the outward voyage. And that would be followed by a three-year tour in orbit around the giant planet. They would be a long way from home: even radio signals, travelling at the speed of light, would take at least 71 minutes to reach Earth.

Mission-experienced

The sixth man, Captain Mason, sat quietly re-reading a telefax from his family. He looked up at the others as they fidgeted in their emergency suits. After six months' intensive training together they had become a close-knit group. All had been selected as much for their compatibility as for their technical expertise and experience.

Olga Belinski, his number two, had also been on a run to Jupiter, as had Tom Lawrence, the nuclear engineer. Gary Richards and Don Maxwell were the mission's science officers, and had both served with distinction in the Martian sector. Suki Yasunari, the communications officer, had worked in the difficult conditions of

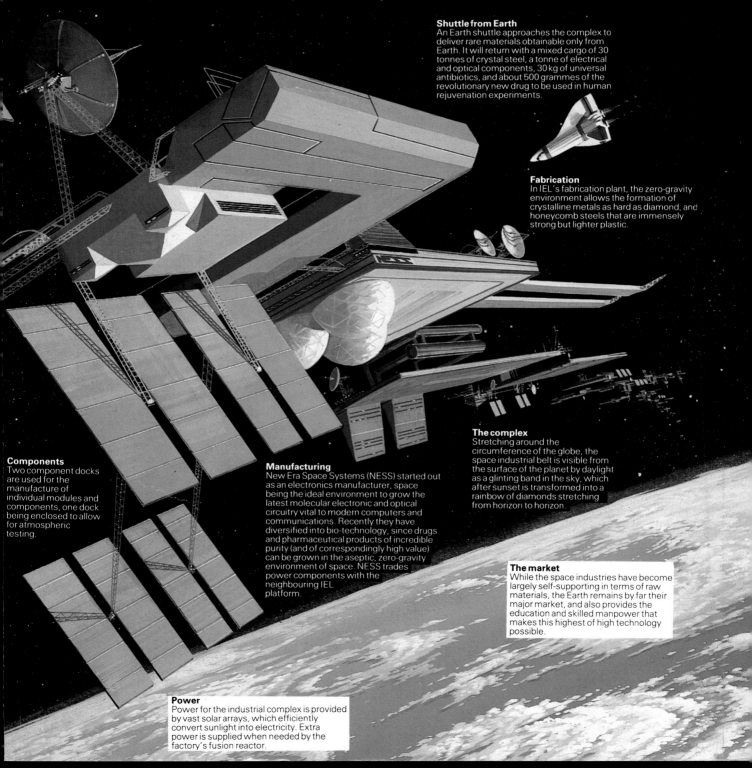

Shuttle from Earth
An Earth shuttle approaches the complex to deliver rare materials obtainable only from Earth. It will return with a mixed cargo of 30 tonnes of crystal steel, a tonne of electrical and optical components, 30 kg of universal antibiotics, and about 500 grammes of the revolutionary new drug to be used in human rejuvenation experiments.

Fabrication
In IEL's fabrication plant, the zero-gravity environment allows the formation of crystalline metals as hard as diamond, and honeycomb steels that are immensely strong but lighter plastic.

Components
Two component docks are used for the manufacture of individual modules and components, one dock being enclosed to allow for atmospheric testing.

Manufacturing
New Era Space Systems (NESS) started out as an electronics manufacturer, space being the ideal environment to grow the latest molecular electronic and optical circuitry vital to modern computers and communications. Recently they have diversified into bio-technology, since drugs and pharmaceutical products of incredible purity (and of correspondingly high value) can be grown in the aseptic, zero-gravity environment of space. NESS trades power components with the neighbouring IEL platform.

The complex
Stretching around the circumference of the globe, the space industrial belt is visible from the surface of the planet by daylight as a glinting band in the sky, which after sunset is transformed into a rainbow of diamonds stretching from horizon to horizon.

The market
While the space industries have become largely self-supporting in terms of raw materials, the Earth remains by far their major market, and also provides the education and skilled manpower that makes this highest of high technology possible.

Power
Power for the industrial complex is provided by vast solar arrays, which efficiently convert sunlight into electricity. Extra power is supplied when needed by the factory's fusion reactor.

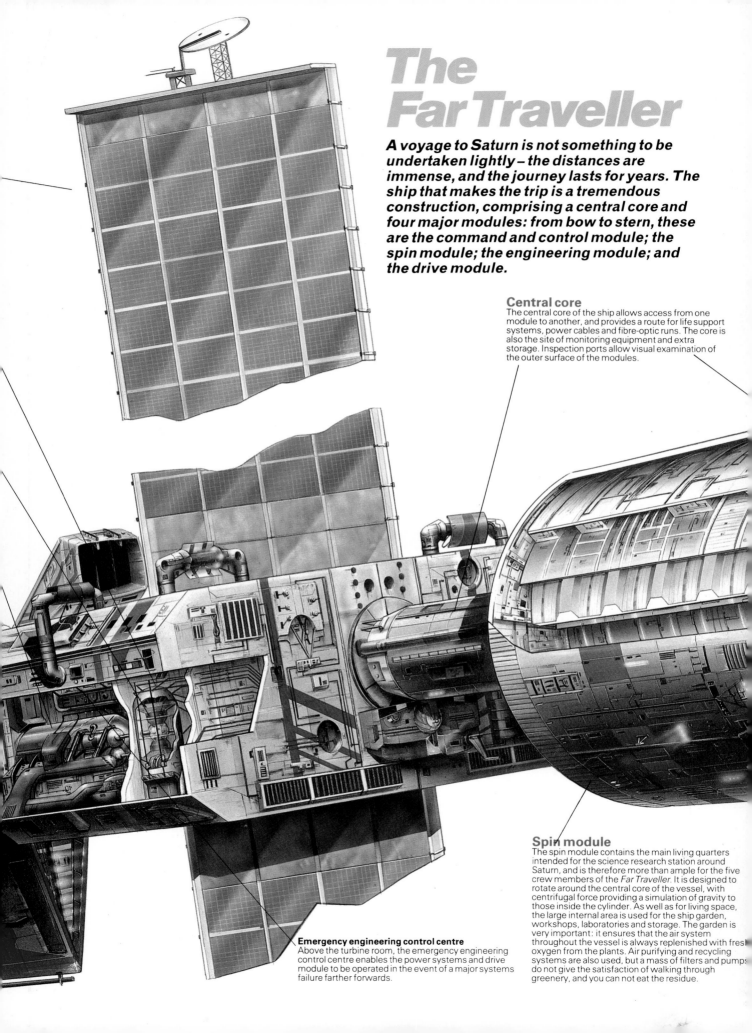

The Far Traveller

A voyage to Saturn is not something to be undertaken lightly – the distances are immense, and the journey lasts for years. The ship that makes the trip is a tremendous construction, comprising a central core and four major modules: from bow to stern, these are the command and control module; the spin module; the engineering module; and the drive module.

Central core
The central core of the ship allows access from one module to another, and provides a route for life support systems, power cables and fibre-optic runs. The core is also the site of monitoring equipment and extra storage. Inspection ports allow visual examination of the outer surface of the modules.

Emergency engineering control centre
Above the turbine room, the emergency engineering control centre enables the power systems and drive module to be operated in the event of a major systems failure farther forwards.

Spin module
The spin module contains the main living quarters intended for the science research station around Saturn, and is therefore more than ample for the five crew members of the *Far Traveller*. It is designed to rotate around the central core of the vessel, with centrifugal force providing a simulation of gravity to those inside the cylinder. As well as for living space, the large internal area is used for the ship garden, workshops, laboratories and storage. The garden is very important: it ensures that the air system throughout the vessel is always replenished with fresh oxygen from the plants. Air purifying and recycling systems are also used, but a mass of filters and pumps do not give the satisfaction of walking through greenery, and you can not eat the residue.

was waiting for them, drifting in the zero gravity. As soon as the lock was sealed he led them, half walking, half floating, from handhold to handhold along a succession of corridors.

They reached the entrance to the living and working quarters and waited for the doors to synchronise between this next section and the support systems area they were in.

Into the pressure vault

A light flashed as the door slid open, and they stepped through into a spacious pressure vault. A variety of spacesuits hung from the walls, including hard torso suits for outside work and radiation outfits. Mason sincerely hoped the latter would not be needed. Only a near-disaster would call for anyone to enter the reactor chamber.

Once the pressures had equalised, the group stripped off their bulky emergency suits and stepped into the accommodation module. It was vast, and their guided tour took nearly three hours to complete. Although it was to be their home for a long time, its ultimate purpose was to provide a new, larger base and research facility for the team already studying Saturn's rings. The ship already stationed there would be their transport back to Earth. The tour over, they spent the next few hours stowing their gear, preparing and eating a well-deserved meal, and exploring the module.

« Is there life on other planets? Titan, one of the moons of Saturn, could be the place to start looking – its atmosphere is similar to Earth's at the time of its formation »

For the next three weeks the new crew underwent further exhaustive training in the ship's systems. They had to be familiar with every aspect of the craft as, once they were more than a few weeks out, they would be too far from the interplanetary shipping lanes to call on any nearby space traffic for emergency help. By the end of this indoctrination period they were all mentally and physically tired, and the prospect of the slow, tedious months and years ahead was positively attractive.

Even the escape capsules had been launched and flight tested, although the crew knew that the life support systems of these tiny craft would last no more than nine months. If anything went drastically wrong while they were at any greater flying time from either end of their voyage, they were doomed.

A simple launch ceremony and televised message from the Central Committee of the Terran Space Agency marked the official start of their six-year journey. Then the crew took up their positions, and a fleet of chubby space tugs locked on to the giant ship to ease it away from the yards. The ship would be towed for two weeks before its nuclear pulse engines could be fully activated. This would both comply with Earth's strict nuclear contamination laws and, at the same time, build up sufficient speed to swing the vessel free from Earth's gravitational field and out into interplanetary space.

The stars began to drift as the tugs heaved at the ship's enormous mass. They were on their way.

Engineering module
The engineering module contains the machinery that will power the ship during the long years of the voyage. Primary power generators are the vast solar arrays, converting sunlight directly into electricity, and the fusion-powered, liquid metal/steam turbine generator systems. Also in the engineering module is the emergency engine control centre and the various engineering workshops.

Solar arrays
The huge solar arrays, spreading like wings from the engineering module, are made from semi-conducting materials such as silicon and gallium arsenide; these can absorb light, which induces an electric current. Large arrays of solar collecters can glean considerable power from the sun.

Fibre optics cables
Fibre optics are used throughout the ship; these are lines which conduct light. These cables direct the intense power of laser light into the fusion chamber of the reactor.

Fusion reactor
The heat for the power system is generated by a laser-fired deuterium/tritium fusion reactor. The heat is carried by liquid sodium to a heat exchanger, where it heats water into steam, which in turn drives the turbines. After use, the steam is piped through to the cooling fin, where it condenses. It is then pumped back to the generator room for a new cycle. A major advantage in using this type of reactor is that one of its waste products is tritium, which is essential to the operation of the main drive.

Turbine room
The turbine room is basically a small electric power station, its steam turbines turning an electrical generator.

Deuterium tank
The empty deuterium tank is jettisoned. On Earth, only about one in every 7,000 atoms of hydrogen are deuterium, but even one cubic metre of seawater could yield enough of the isotope to provide as much energy as several hundred tons of coal being burned in a power station.

Drive module
This contains the main propulsion system, developed from that of the abortive 'Daedalus' unmanned starship. Two forms of hydrogen – deuterium and tritium – are stored in pressurized pellet form and, fused together under the influence of a battery of electron-beam powered lasers, the two isotopes of hydrogen form helium, with part of their mass converted to pure energy. This provides the motive power for the massive ship.

Tritium container
Tritium does not exist naturally, but can be manufactured by bombarding the metal lithium with neutrons; or it can be made from deuterium in another kind of fusion reactor. It is stored in spherical containers around the fusion engine.

Motor nozzles
Much of the space inside the drive module is taken up by the nozzles of the fusion motors. Around these huge mounts, energy is drawn off from the plasma, as the unimaginably hot product of fusion reaction is called.

Cooling fin
The main drive is in operation for a minute portion of the journey, so the drive module is a good location for the cooling fin. This radiates away excess heat that comes from a variety of sources, ranging from the direct action of the sun, unfiltered by any atmosphere, to the heat generated by the fusion reactor in the engineering module.

the asteroid belt for nine years before joining the TSA's survey division.

At last they were instructed to board. Mason led them along the walkway to the shuttle. Once aboard, they slipped one by one into their seats, exchanging greetings with the vessel's crew, who were running through the final stages of countdown. Twenty minutes later the carrier's main drives fired, and the whole assembly took off, climbing steeply. At 50,000 metres the orbiter's main engines took over, and orbiter and launcher separated, one to return to base and the other to continue upwards out of the atmosphere.

As the shuttle drifted through space in the last stages of its journey to the spaceship, the younger members of the crew were staring at the distant Earth's bright, cloud-streaked surface. Mason was gazing in the other direction: the spectacle of the orbital shipyards and industrial units fascinated him more. Then he saw their own destination. He hadn't noticed it at first, as they were approaching from the dark side of the huge craft. It looked like a slender black patch in the curtain of stars until the shuttle began to curve round on its approach course. The nearer they drew, the more stars were blotted out, until the ship filled the entire viewport.

Docking manoeuvre

The shuttle skipper began the delicate manoeuvre that would connect them with one of the ship's docking bays. There was a soft thump as they locked on and a hiss as pressures equalised. The pilot opened the airlock and gave them an OK signal, and they began to push themselves through the narrow opening in the shuttle's hull. A shipyard technician

The Orbital Industrial Complex

Raw materials
The Moon is a source of much of the raw material used in the belt. With only one sixth the surface gravity of the Earth, it requires considerably less energy to get oxygen and ore-bearing rocks into space.

Command module
The command, control and communication systems of the command module have been thoroughly examined, and back-up systems have been tested. Throughout the voyage the ship will be in contact with mission control on Earth, although the further they go the longer the signals will take to be exchanged. Continuous telemetry will provide thorough details of the ship's progress and performance.

Spin module
The spin module is tested to ensure it will function effectively through the years of the flight. It is designed to rotate three times per minute, providing a simulation of half earth gravity to someone standing on the inner surface. This is important, as long periods of weightlessness can cause irreparable damage to the strength of the human bone structure.

The ship
The completed Saturn Traveller is in a parking orbit 1 km from the dockyard. Final checks to the drive are being completed, and the fuel tanks atop the drive module are ready to receive their Deuterium/Tritium fuel pellets.

Engineering module
While in the inner parts of the Solar System, most ship's power will be provided by the 200-metre long solar panels. In the outer reaches, however, more power will come from the nuclear reactor in the engineering module immediately forward of the drive module.

Loran/GPS satellite navigation system
The navigator uses these to fix his position electronically to within a few metres, making use of the navigation satellites high in orbit above the Earth.

Starboard navigation radar
One of a pair carried by *Challenger II*; radar is essential if you want to make a high-speed run at night or in fog, as it gives warning of icebergs or ships in your course. Unfortunately it is of little use when confronted by a whale surfacing 50 metres ahead of your speeding boat!

Navigator's station
This is equipped with every electronic navigation and communication device imaginable. In addition to the radios, radar and satellite plotting systems, the *Challenger II* carried a computerised electronic chart and a radio telex link with the 'Virgin Challenge' office in London.

Twin helms
The boat is controlled a engine monitored from An autopilot with twin gyrocompasses is fitte just in case, *Challenger* four magnetic compass aboard during the cross

Welded aluminium hul
This is a compromise be the need for high speed a strength and high fuel loa required to make a high-s crossing of the Atlantic.

Crew seats
These are designed to be comfortable during the long high-speed run through the swells of the North Atlantic.

Gearbox
This is housed in a lightweight aluminium case fitted aft of the MTU engine. Although able to reverse the propeller, the main function of the box is to reduce the 2,100 revolutions per minute of the engine to less than 200 rpm at the propeller shaft.

Starboard MTU V-12 diesel engine
The engine produces 1440 kW at 2100 rpm. There are over 7,500 Series 396 engines in service, mainly in ships or trains. *Challenger*'s pair were standard production units.

Main fuel tanks
Built in aircraft fashion, their rubber outer walls contain a honeycomb foam that actually holds the fuel. There are four tanks, each with a capacity of 3.13 tonnes, giving *Challenger II* a range of about 1528 km at 45 knots.

The American holders of the Hales Trophy, awarded for the fastest crossing of the Atlantic, say that it was meant for full-sized ships. Argument continues as to whether Challenger was eligible, but the fact remains that she made the fastest-ever crossing.

another attempt on the record. This time the boat was to be a mono-hull, drawn up by a pioneering boat and aircraft designer called Sonny Levi. *Virgin Atlantic Challenger II* was 21.9 metres long, a sleek dart of aluminium with the same engines as had been used in its predecessor. The new boat could plane at 50 knots, and skidded off a wave in trials, breaking the leg of one would-be crew-man. The six-man crew for the attempt was Branson; Chay Blyth, a well-known marine record-breaker; Dag Pike, a journalist and experienced sailor; Steve Ridgway, who had helped build the first *Challenger*; Peter Macann, a BBC presenter; and

Eckhard Rastig, a marine engineer who installed the engines.

Trailing her huge wake, churned white by massive twin screws, *Challenger II* pounded away from the starting line at 10.03 am on 27 June 1986. Negotiating whales, icebergs and fog, *Challenger II* thundered to her first refuelling point, and onwards. At night and in fog they were powering 'blind', with the aid of satellite navigation, electronic charting and twin radar units. Then, at the second refuelling stage, water entered the fuel filters, along with 12.5 tonnes of fuel. Agonising hours of draining, refilling and filter-changing ensued, eroding the precious time left to beat the 34-year-old record. Keeping the engines running became a nightmare, and spare filters were even dropped from an RAF Nimrod. Heavy weather on the third day

slowed *Challenger II* down, but the last 650 km were raced in good con ditions, flat out, and the exhauste but exhilarated crew stormed ove the finishing line 3 days, 8 hours an 31 minutes after crossing the start They had beaten the *United States* record by 2 hours 9 minutes.

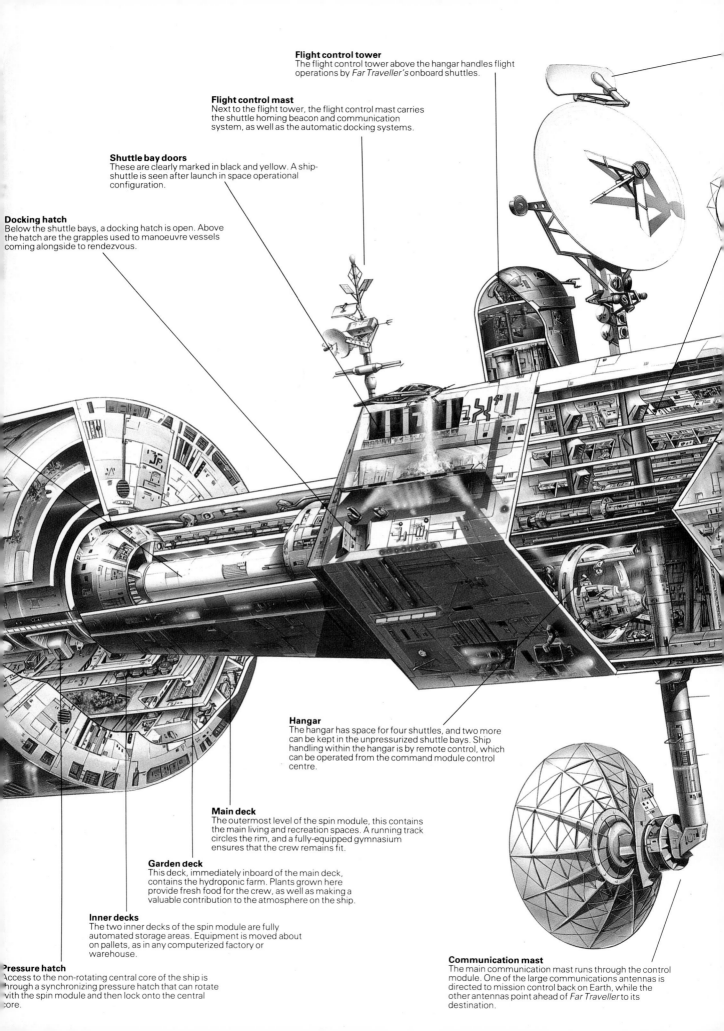

Flight control tower
The flight control tower above the hangar handles flight operations by *Far Traveller's* onboard shuttles.

Flight control mast
Next to the flight tower, the flight control mast carries the shuttle homing beacon and communication system, as well as the automatic docking systems.

Shuttle bay doors
These are clearly marked in black and yellow. A ship-shuttle is seen after launch in space operational configuration.

Docking hatch
Below the shuttle bays, a docking hatch is open. Above the hatch are the grapples used to manoeuvre vessels coming alongside to rendezvous.

Hangar
The hangar has space for four shuttles, and two more can be kept in the unpressurized shuttle bays. Ship handling within the hangar is by remote control, which can be operated from the command module control centre.

Main deck
The outermost level of the spin module, this contains the main living and recreation spaces. A running track circles the rim, and a fully-equipped gymnasium ensures that the crew remains fit.

Garden deck
This deck, immediately inboard of the main deck, contains the hydroponic farm. Plants grown here provide fresh food for the crew, as well as making a valuable contribution to the atmosphere on the ship.

Inner decks
The two inner decks of the spin module are fully automated storage areas. Equipment is moved about on pallets, as in any computerized factory or warehouse.

Pressure hatch
Access to the non-rotating central core of the ship is through a synchronizing pressure hatch that can rotate with the spin module and then lock onto the central core.

Communication mast
The main communication mast runs through the control module. One of the large communications antennas is directed to mission control back on Earth, while the other antennas point ahead of *Far Traveller* to its destination.

Radar system
On top of one of the Saturn communication dishes, a 3D radar system scans the space ahead of the ship for space debris directly in the ship's path.

Decks
Above the hangar are three laboratory and workshop decks, with the communication and computer deck immediately above. Crew access to the main communication mast and to the flight control tower is through this deck.

Watch-keeping facilities
An autogalley and wardroom behind the control deck have recreational and cleansing facilities for watch-keepers.

Escape pods
The emergency escape pods have with launching tubes port and starboard. The pods are equipped with solar-powered distress beacons and hibernation units.

Food storage areas
Frozen foods are placed aft of the concentrates. Inspection pods run suspended from a rail through the stores although, as in the rest of the ship, operation is largely automatic.

Control centre
The rear section of the control deck is devoted to astrogation and communications, while the central area has the ship monitoring systems and computer centre links. Forward is the flight deck.

Control module
The control module has a number of different functions, and like the rest of the ship is highly automated. Flight deck, control room, electronics, life support, astrogation, hangar and consumables storage are all located in this section, together with communications and diverse science projects.

Life support system
Life support machinery is largely concerned with the recycling of oxygen, heating, lighting, and the conservation of water. The system is largely automatic and is monitored from the control centre, but there is a life support engineering control for use when adjustments have to be made.

Attitude jets
These are located bow and stern, and are used to position the ship before the main drive fires.

Mini-shuttles
These are primarily for use in space but can operate in atmosphere; they have been designed for use on Titan, the moon of Saturn. Titan has a thick atmosphere of nitrogen and methane, similar in many respects to that of Earth before the evolution of life.

Chemical storage tanks
Liquid oxygen and materials used in the recycling process are stored here. These are themselves recycled, with the small amount of totally useless waste eventually being jettisoned.

Atlantic Challenge

Radio aerials
These are connected to the boat's high-frequency and VHF communication systems. Walkie-talkies were also carried, as they were vital during the refuelling stops in mid-ocean.

Crew toilet
This and the rest of the accommodation is in the cockpit; the normal accommodation area was not used.

Engine air intake
Huge quantities of air are required by the big V-12 engines, so their intakes are filtered to stop sea water getting into the system. As it happened, the problem was to stop water coming in from aboard the refuelling ships!

For almost 150 years commercial passenger ships have battled for the honour of holding the Blue Riband, the award made for the fastest sea-crossing between the Ambrose Light Tower off New York, and the Bishop Rock Lighthouse off the Scilly Isles. In 1933 another prize, the Hales Trophy, an ornate, 1.2-metre, 45-kg mass of gold-plated silver, was added to the competition. The record for the 2,949 nautical miles (5461 km) was continually broken over the years, from 15 days in 1838, to 3 days, 10 hours and 40 minutes in 1952, when the honours were won by the 52,000-tonne *United States*. And there the record stuck.

Liferafts
Essential equipment on any long voyage, but especially to Branson and his crew: after all, they had to take to a raft after *Challenger I* sank the year before.

Spare propellers
Mounted on a rack at the stern, these can be lowered easily into the water and fixed in position by swimmers.

Semi-circular rudders
These fit around the propellers and direct the flow of water from the screws, increasing their propulsive efficiency.

Levi propulsion units
The exhaust, propeller and rudder are combined in a single unit. The propellers are surface-piercing, which means that at speed they break the surface of the water – more effective at the speeds *Challenger II* can attain.

CHRIS BAKER

Challenger pounds over the waves. She is capable of nearly 53 knots – over 60 mph – when lightly loaded; even with well-cushioned seats, the crew of the Challenger took a considerable buffeting at that speed.

In 1985 Richard Branson, multi-millionaire boss of the Virgin group of companies, made his first attempt to win the trophy for Britain – and to grab a sizeable hunk of publicity for his business enterprises at the same time. The result of five years' planning, Branson's *Challenger I* was a space-age catamaran in aluminium, capable of close to 50 knots under the 4,000 hp thrust of its twin diesel engines. *Challenger I* set off on 10 August 1985 from the Ambrose Light Tower and seemed sure of the record until floating debris wrecked the boat some 480 km short of the finishing line.

A less determined – and less wealthy – man might have given up but, remarking that he had 'a little piece of unfinished business', Richard Branson set about planning

ROUTE OF THE CHALLENGE

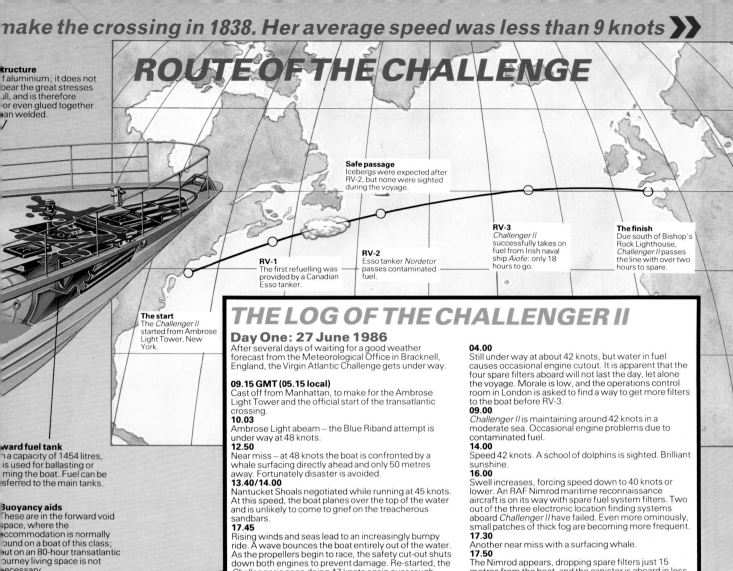

Structure
f aluminium; it does not
bear the great stresses
ull, and is therefore
or even glued together
an welded.

Safe passage
Icebergs were expected after RV-2, but none were sighted during the voyage.

RV-1
The first refuelling was provided by a Canadian Esso tanker.

RV-2
Esso tanker *Nordetor* passes contaminated fuel.

RV-3
Challenger II successfully takes on fuel from Irish naval ship *Aiofe*: only 18 hours to go.

The finish
Due south of Bishop's Rock Lighthouse, *Challenger II* passes the line with over two hours to spare.

The start
The *Challenger II* started from Ambrose Light Tower, New York.

ward fuel tank
a capacity of 1454 litres,
is used for ballasting or
ming the boat. Fuel can be
sferred to the main tanks.

Buoyancy aids
These are in the forward void
space, where the
accommodation is normally
found on a boat of this class;
but on an 80-hour transatlantic
journey living space is not
necessary.

« *Challenger II's*
average speed was
6.54 knots, beating the
me set in 1952 by the
ner United States,
hich managed a
6.59 knot average »»

THE LOG OF THE CHALLENGER II

Day One: 27 June 1986
After several days of waiting for a good weather forecast from the Meteorological Office in Bracknell, England, the Virgin Atlantic Challenge gets under way.
09.15 GMT (05.15 local)
Cast off from Manhattan, to make for the Ambrose Light Tower and the official start of the transatlantic crossing.
10.03
Ambrose Light abeam – the Blue Riband attempt is under way at 48 knots.
12.50
Near miss – at 48 knots the boat is confronted by a whale surfacing directly ahead and only 50 metres away. Fortunately disaster is avoided.
13.40/14.00
Nantucket Shoals negotiated while running at 45 knots. At this speed, the boat planes over the top of the water and is unlikely to come to grief on the treacherous sandbars.
17.45
Rising winds and seas lead to an increasingly bumpy ride. A wave bounces the boat entirely out of the water. As the propellers begin to race, the safety cut-out shuts down both engines to prevent damage. Re-started, the *Challenger* is soon doing 47 knots again over rough seas.
20.30
RV-1 (Rendezvous 1) is reached some 35 km off Halifax, Nova Scotia. Refuelling begins from a Canadian offshore support vessel but, in the tricky conditions, the operation takes close to an hour and a half.
21.50
Set off for RV-2, off the eastern tip of Newfoundland. Speed through a moderate sea and swell is approximately 42 knots.
24.00
2 hrs 40 mins ahead of record schedule. Speed through the night is some 42 knots. Lookout is possible only by radar.

Day Two: 28 June 1986
09.00
After 24 hours at sea *Challenger II* passes through 'Iceberg Alley', but none of the predicted bergs are sighted.
12.00
Brightening weather and clearing seas raise the average speed to 47 knots. Fog patches begin to appear.
13.00/15.00
Despite thickening fog patches all the way to RV-2, *Challenger II* ties up alongside the oil supply ship *Nordetor* by 15.00.
15.40
Fuelling complete – but engines do not start after casting off.
16.00
Fuel filters are found to be full of water – the 12.5 tonnes of fuel taken aboard is contaminated. Later it is calculated that at least 4 tonnes of water had mixed with the fuel oil.
16.00/23.00
Tanks drained and refilled – but residue of fuel/water mixture continually foils efforts to start engines.

Day Three: 29 June 1986
00.15
Finally finish flushing out fuel tanks and refilling. Cast off from RV-2 and set off into moderate swell at 42 knots.

04.00
Still under way at about 42 knots, but water in fuel causes occasional engine cutout. It is apparent that the four spare filters aboard will not last the day, let alone the voyage. Morale is low, and the operations control room in London is asked to find a way to get more filters to the boat before RV-3.
09.00
Challenger II is maintaining around 42 knots in a moderate sea. Occasional engine problems due to contaminated fuel.
14.00
Speed 42 knots. A school of dolphins is sighted. Brilliant sunshine.
16.00
Swell increases, forcing speed down to 40 knots or lower. An RAF Nimrod maritime reconnaissance aircraft is on its way with spare fuel system filters. Two out of the three electronic location finding systems aboard *Challenger II* have failed. Even more ominously, small patches of thick fog are becoming more frequent.
17.30
Another near miss with a surfacing whale.
17.50
The Nimrod appears, dropping spare filters just 15 metres from the boat, and the canister is aboard in less than half a minute. *Challenger II* sets off for RV-3, the Irish Navy's patrol ship *Aiofe*.
23.45
RV-3 and, in spite of heavier weather, refuelling is uneventful. Within half an hour the boat is proceeding on the final stage of the transatlantic journey.

Day Four: 30 June 1986
00.15
Challenger II makes slight detour to avoid weather front and allow higher speeds to be maintained.
08.30
Missed worst of weather, but conditions are still difficult. Speed drops to 38 knots.
11.00
Weather has improved dramatically. The last 650-km run can be made at 48 or even 50 knots as the fuel load is consumed.
16.00
At 50 knots, the first ship not related to the attempt on the record is sighted. Conditions fine and clear. More whales sighted, but none on collision courses.
18.10
First radar contact with Bishop's Rock Lighthouse. Good weather has given way to heavy rain squalls, but the boat maintains a speed of 50 knots. To complete the official Blue Riband route the *Challenger II* has to cross a finish line stretching due south of the light.
18.30 (approx.)
Bishop's Rock sighted visually at about three miles. Several helicopters escort the boat in as she speeds, still making 50 knots, towards the end of her voyage.
18.34 GMT
Challenger II passes due south of Bishop's Rock Lighthouse, having completed the transatlantic crossing in 3 days, 8 hours and 31 minutes. She thus beats the record held by the liner United States since 1952 by 2 hours and 9 minutes.
19.20
Surrounded by an armada of welcoming craft, *Challenger II* makes fast alongside the harbour of St. Mary's in the Scillies.

SA80: REVOLUTIONARY NEW ARMY RIFLE

The Corporal cradled the rifle across his forearms as he crawled through the undergrowth, first one elbow, then the other, hauling himself along so slowly that no branches or grasses quivered above him. It was just as well, he thought, that this new rifle was a good 30 cm shorter than the old one, otherwise the muzzle would surely have caught in a bush and given him away. It was lighter, too, and a lot less tiring to crawl with.

He edged his way forward, finally stopping underneath a bush on the hillside; never hide on the crest, that's where they always look, hide off on the side of the hill. He scanned the hedge, about 200 metres away, carefully, end to end. There was a

sniper in there somewhere, had to be. Someone had wounded two of his men back there. The only place they could have been was that hedge.

Slowly he eased the rifle forward; no need to worry about sunlight reflecting off it, the metal was dull and the rest of the gun was plastic. He looked through the optical sight, and the hedgerow leapt towards him, magnified four times. He scanned again, slowly. There. Was it? Yes. A movement of a bush. He watched carefully. Easy to hold this rifle, nice and comfortable; I can keep it against my shoulder all day if I have to. There he is. I can just see his

helmet. Just put your head up matey... You won't, eh? Never mind, this bullet will go through your helmet at this range; not like the old 5.56 bullet we tried with the M16 rifle. The corporal laid the pointer on the painted badge on the side of the helmet and squeezed the trigger. There was a sharp crack, and the helmet flopped sideways. Looks like that's done it; good job there's so little recoil with the 5.56 mm cartridge. I'm back on aim, might as well give him another for good measure. Crack. That's it. Back to the lads.

Sight
Unusually, all operational versions of the SA80 will be fitted with optical sights. The SUSAT sight has a ×4 magnification, and allows soldiers to be effective even in poor light. It can also be used for observation.

Piston
The gas vented from the barrel is fed into a gas cylinder, which drives against a piston running along the top of the weapon. The piston is forced backwards against the firing mechanism.

Barrel
The Bullpup design has a number of advantages in a combat rifle. The rearrangement of trigger and magazine has done away with the need for a normal butt, and overall the weapon is about 40 per cent shorter than its predecessors.

Flash eliminator
This device allows escaping gases to expand and cool, preventing the formation of flash and flame.

Gas vent and cylinder
About half way up the barrel, a small hole in the top taps off a little of the gas caused by the explosion of the cartridge. The energy of this hot, rapidly-expanding mass is so great that a relatively tiny amount is all that is required to work the rifle.

Trigger
The trigger of all weapons releases a catch holding the hammer back. Normally, the trigger is close to the hammer, but the SA80 is different: it is a 'Bullpup' design, meaning that the magazine is *behind* the trigger rather than in front, so the trigger connects with the hammer by means of a long trigger bar.

Magazine
The NATO standard 5.56mm ammunition is carried in a 30-round box magazine that is interchangeable with the American Armalite.

❮❮ The SA 80 (Small Arms 80) family will equip the entire British Army with personal weapons for less than the cost of a single frigate or half a squadron of Tornado aircraft ❯❯

The British Army's new rifle has come as quite a shock to the old traditionalists. It looks decidedly odd; it's short, the magazine is placed *behind* the trigger, and there is a telescopic sight for every user, not just for the sniper. It just doesn't look like a battle rifle!

But a lot of thought has gone into its design. The old weapons fired high-power bullets, effective to a range of 100 metres or more, yet most soldiers

fight at ranges of 30 metres or less, so a lighter bullet would make sense; this also means that soldiers can carry more ammo into battle. Modern soldiers also spend far more time inside their cramped armoured personnel carriers or in helicopters, so a shorter weapon would make life far more comfortable. And the telescopic sight is not designed to turn everyone into a sharpshooter but to help when fighting in low visibility. War, unlike

cricket, does not stop for bad light!

The SA 80 is much cheaper to make than the previous Army rifle, the SLR, which means it's popular with the politicians. Most of the weapon is nylon/plastic and pressed steel, formed in moulds, whereas the working parts of older rifles were usually expensively machined from solid metal, a slow and expensive process.

Of course, the most obvious difference between the SA 80 and most

FIRING POSITIONS

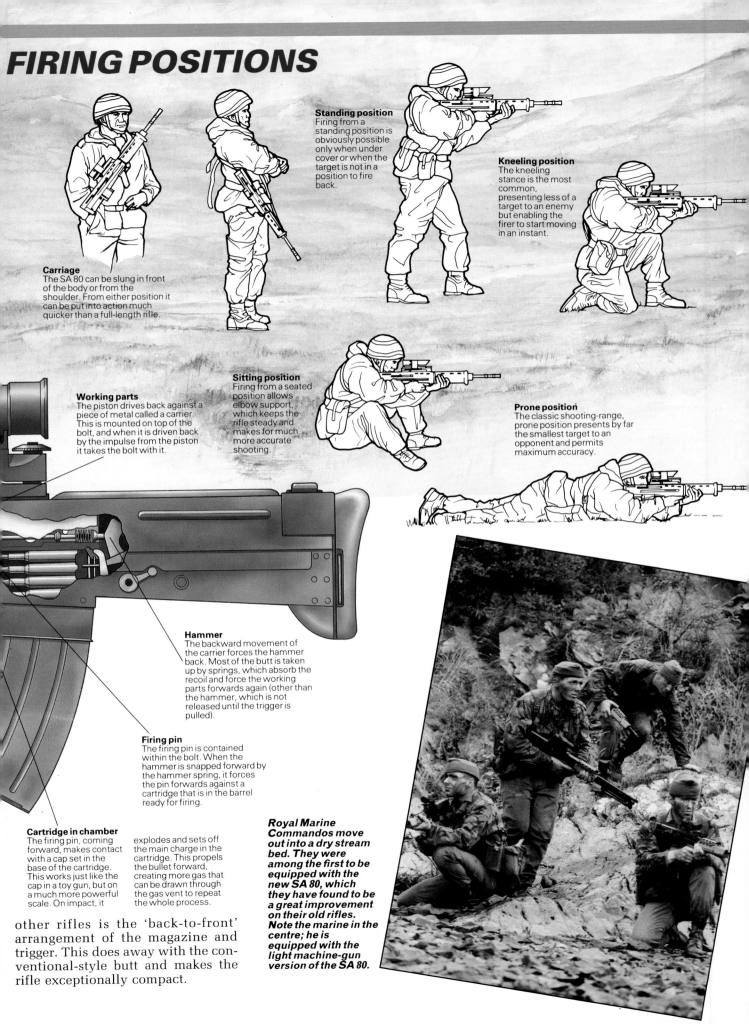

Carriage
The SA 80 can be slung in front of the body or from the shoulder. From either position it can be put into action much quicker than a full-length rifle.

Standing position
Firing from a standing position is obviously possible only when under cover or when the target is not in a position to fire back.

Kneeling position
The kneeling stance is the most common, presenting less of a target to an enemy but enabling the firer to start moving in an instant.

Working parts
The piston drives back against a piece of metal called a carrier. This is mounted on top of the bolt, and when it is driven back by the impulse from the piston it takes the bolt with it.

Sitting position
Firing from a seated position allows elbow support, which keeps the rifle steady and makes for much more accurate shooting.

Prone position
The classic shooting-range, prone position presents by far the smallest target to an opponent and permits maximum accuracy.

Hammer
The backward movement of the carrier forces the hammer back. Most of the butt is taken up by springs, which absorb the recoil and force the working parts forwards again (other than the hammer, which is not released until the trigger is pulled).

Firing pin
The firing pin is contained within the bolt. When the hammer is snapped forward by the hammer spring, it forces the pin forwards against a cartridge that is in the barrel ready for firing.

Cartridge in chamber
The firing pin, coming forward, makes contact with a cap set in the base of the cartridge. This works just like the cap in a toy gun, but on a much more powerful scale. On impact, it explodes and sets off the main charge in the cartridge. This propels the bullet forward, creating more gas that can be drawn through the gas vent to repeat the whole process.

Royal Marine Commandos move out into a dry stream bed. They were among the first to be equipped with the new SA 80, which they have found to be a great improvement on their old rifles. Note the marine in the centre; he is equipped with the light machine-gun version of the SA 80.

other rifles is the 'back-to-front' arrangement of the magazine and trigger. This does away with the conventional-style butt and makes the rifle exceptionally compact.

NATURE'S NIGHT FIGHTERS

The summer night sky is full of insects softly buzzing about their business. But other creatures are flying tonight: mammals whose 'hands' have evolved into wings, whose noses have become radar-transmitters and who can fly in darkness, sensing obstructions as thin as hairs.

A single greater horseshoe bat is caught in flight in its search for food. Its wings are the equivalent of our hands, with their membranes of skin stretched between the elongated fingers.

Bats are famous for finding their food by sending out high-pitched squeaks and homing in on the source of the echo. It's ingenious, perfectly suited to a night hunter, and it seems so simple. But look at it from the bat's point of view, and you soon realise that the bat's detector system – known as echo-location – is in fact subtle, complex, and even more ingenious than you first thought.

A bat can cruise at 30 mph. Somehow, at this pace, it has to detect a speeding 5 mm-long midge, locate it exactly, and close in for the kill. This is rather like trying to score a bullseye on a moving dartboard while leaning from an express train.

To succeed, the bat *aims* its squeak. Otherwise the sound would be bouncing off all kinds of things, all over the place, which would be not only confusing but useless to a hungry bat. So bats use their mouths or, if they're the kind that squeaks through their nose, a 'noseleaf' to focus the sound into a narrow beam.

The bat also has to sort out the echoes it hears. At close quarters a squeak that lasts only 0.05 seconds can still be sounding as its echo returns. At the same time it may be picking up the squeaks of other bats. And finally the bat has to be able to tell the echo of something worth catching from that of something inedible.

Food for thought

Bats solve these problems by *interpreting* what they hear, and by adjusting the frequency of their sounds so that they can distinguish their own signals of those of other bats nearby. At the same time the bat will be listening out for an insect's distinctive, individual buzz. A bat's echo-location system is often said to be like radar, but it's more like a radar set, technician, computer and high-speed air-traffic controller all rolled into one.

Once locked on to its prey, the bat traps it in its teeth. It may miss – but then it'll scoop the insect up in its wingtip, like a cricketer taking a difficult catch. Sometimes the prey is caught in the pocket formed by the bat's tail membrane. The bat will pick it out of here with its mouth – all while on the wing.

« Some bats, using echo-location, can detect presence of wire as fine as human hair »

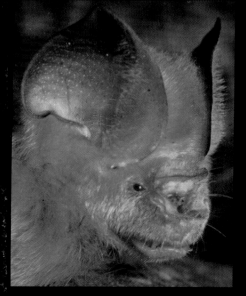

Well, if you had a radar transmitter for a nose you'd look strange too. Bats emit squeaks from their mouths or, as in the case of this African leaf-nosed bat, from their noses. Those that use their nostrils have often evolved complicated and sometimes hideous folds and flaps to help amplify and resolve the squeak.

NIGHT OPERATIONS

Bats are far better equipped than fighter pilots for their deadly night operations. They can use a sort of radar over a huge range of frequencies, sort out their own returns from those of other bats, and visualise what prey is being intercepted.

On patrol
Bats, while cruising, emit pulses at the rate of about two per second. The emissions are well spread out, scanning the environment and ground to produce a generalised picture in their minds.

Terrain avoidance
Once the bat gets a radar return, it steps up its transmissions and reduces beam width. This improves the resolution of the picture in the bat's mind and enables it to fly through complicated terrain such as woodland.

Different tactics
Different bats feed in different ways. Some specialise in catching big moths, carrying their victims off to some convenient roost and tearing them apart. Others scoop tiny swarms of insects from the air by the mouthful, or snap at individuals as they flash through an airborne crowd of insects.

Night fishing
Fisherman bats patrol ponds and rivers at night, picking up fish and frogs from the surface.

Sharp end
Bats' teeth are needle-sharp to rip off the tough outer covering of their favourite food.

Blood-drinking
Vampire bats bite the ankles of cattle and other animals and lap up the dripping blood.

Desert predator
Some desert bats cruise close to the ground to swoop on scorpions and large beetles.

In for the kill. The bat will either catch the moth in its mouth or will scoop it up in the flap of skin between its legs and transfer it to its mouth while in flight.

Hanging around
The false vampire bats of Africa hang up and wait for a suitable – that is, big enough – victim to come within range before they pounce.

Season diets
In cool countries, some bats eat voraciously throughout the summer to see them through the insectless months of winter, while others migrate to warmer climes.

Vegan bats
Some bats don't eat insects at all, but live on fruit or the nectar of giant flowers.

Prey ahead
As soon as echoes show the presence of prey, pulse repetition rate increases to 20 to 30 clicks per second. Some bats can now lock on to the buzzing of the insect's wing, while others move their heads from side to side to get a fix using the Doppler (double-pulse) effect.

Close contact
As the bat closes in on the insect, pulses are stepped up to over 100 per second for accurate last-minute detection.

Counter-measures
Some insects react to the bat's piping voice by spiralling away or closing their wings and falling out of the air. A few moths can confuse the attacker by transmitting clicks of their own on similar frequencies to those of the bat's.

WRONG TURN IN THE DESERT

On Friday, 11 June 1965, aircraft of the Egyptian Air Force Rescue Corps found the first bodies, completely dessicated by the ferocious sun, of two men and a woman. The next day they found the other two men. All belonged to the German expatriate community in Cairo. Klaus and Gudrun Boehme had run a shop in the Egyptian capital. Reinhold Rimm, Guenter Wanderscheck and Hans Hauser had worked at an aircraft factory south of the city. With another couple of friends they had set out for what was intended to be an exciting tourist trip west along the Mediterranean coast from Alexandria and through El Alamein. It was a region of great significance to the Germans: the great General Rommel, the Desert Fox, had brilliantly outmanoeuvred the Allied forces there with his crack Afrika Korps, almost succeeding in pushing through to the Egyptian heartland before he was stopped at El Alamein. Many Germans already lay dead in those hostile sands.

Deadly depression

When the seven tourists reached El Alamein, they made a fatal decision. Most of them wanted to drive as far as the Siwa Oasis, deep in the Western Desert and close to the Libyan border. Wheeled traffic for Siwa stays on the coast road as far west as Mersa Matruh, then follows the telephone lines south-west to the Oasis. It's a long way round, but for very good reasons. The direct route from El Alamein to Siwa is shorter on the map, but goes straight through the centre of the Qattra Depression, the most difficult and deadly part of the Western Desert. On Saturday, 5 June, the three aircraft workers and the couple with the shop decided to take the short cut to Siwa. Their two friends turned back. The five adventurers, confident in their two VW cars, set off into the desert.

The Qattara Depression is an 18,000 sq km scar gouged by the wind out of the desert floor. Much of it is below sea level – 134 metres below at the deepest point. The terrain is a vehicle-wrecking mix of soft sand, salt-marsh, rocks and cliffs. Rommel had given it a wide berth, considering it to be impassable even to his tracked

vehicles. The heat in June is savage, killing unwatered life within hours, and instantly mummifying corpses.

The grim record

A film found in Gudrun Boehme's camera chronicled the inevitable tragedy. In one photo Hans Hauser is lying under a car, trying to find out what went wrong. All car doors are open in the attempt to cool the furnace-like interiors. In further shots the men stand around disconsolately, drinking greedily from their plastic canteens. In another shot they are sitting on the ground, the sun is lower. They are still drinking.

In the last stages of dehydration the dry heat pulls the blood to the surface of the skin, where it evaporates, leaving clogging sediments that swiftly block the arteries. Death comes painfully, accompanied by madness, as the circulation system packs up. When Gudrun shot the last exposure on the film, she must have known she was recording the end. The men lie weakly in the sand, which they have pulled up half-heartedly over the bodies to try and ward off the sun.

It had all happened so fast. They were only taking a short cut. It had seemed such a good idea.

In the first picture, the tourists are still alert and on their feet. But one of them is not wearing protective headgear – a bad mistake.

The second one shows two of the men, near the end of their strength, attempting to gain some shelter from the fierce sun by covering themselves with sand.

HOW TO SURVIVE IN THE DESERT

Desert terrains vary from sand and stone to salt and dust. They can be flat or mountainous. What they all have in common is a lack of water. Death comes in several forms. Sandstorms can choke you, strip the flesh from your bones, and bury you. You can even drown in a flash flood if you sleep in a dry watercourse. Most deaths, however, are from dehydration, sun scorching, or a combination of the two. Starvation is rare, because you die swiftly in the desert.

To survive vehicle breakdown in the desert the following basic precautions should be taken.

1 The availability of water dominates all other considerations. Before setting out you should have stowed aboard at the very least five litres per person per day. If you think the journey will take five days, carry enough water for 10 days. If and when your vehicle breaks down, carefully secure the water supply in the shade, then use it as efficiently as possible.

2 As dehydration is the main enemy, water is best eked out by lessening the body's water loss. The nomadic tribesmen of North Africa dress ideally for the desert, in many-layered, loose-flowing garments. Despite the urge to tear clothing off in the heat of the day, stay full clothed, leaving as little skin exposed as possible.

Improvise a head-dress from spare clothing, arranging it if possible to wrap the head and cover the neck. You should be able to draw it across the nose and lower face against a sand-storm. Apply any sun-screen lotions liberally to exposed skin, and keep the eyes shaded as much as possible to avoid sun-blindness.

3 Don't stay in the vehicle, as it swiftly turns into an oven under the fierce sun. If you have the materials, rig a sun shade. You can sleep in the vehicle at night, and it may be useful in case of sand-storm, although these have been known to bury vehicles as large as tanks. If caught in the open during a sandstorm, lie on the ground with your back to the wind, and your face covered.

4 Supplement water supplies if possible. If you can find plants, you can make a vegetation still. Put vegetable materials such as leaves, cactus pulp, fruits or flowers into a large plastic bag. Inflate the big, tie it shut, and put it in the sun. For a solar still, dig a hole about 60 cm deep and 90 cm square. Put a container in the bottom of the tapering hole, and suspend above it a plastic sheet, secured with rocks around the rim of the hole. Weight it in the middle with another rock, so that condensation will drip into the container. Vegetable matter in the hole helps the process.

5 Confine activity to the night-time, and rest in the shade during the day. This will help conserve energy and body moisture.

6 If you hear an aircraft or another vehicle, a fire is a good night-time signal. During the day the smoke from burning oil or from a burning tyre can be seen for a long way. For visible rescuers a signalling mirror can also be seen for miles.

7 Don't leave the vehicle to walk out of your predicament unless there is no other chance of being found. You should already have told the authorities and your friends your destinations and routes, as well as your estimated times of arrival. If you have to attempt to walk out, carry as much water as possible, leave a message and compass bearing at the vehicle, navigate carefully, and travel only when the sun is very low or at night when visibility is good. If you travel in the cool hours, properly dressed, looking after your feet and controlling your perspiration, you could make 30 km a day on five litres of water.

8 Food is the least of your problems, as you can last longer without it than you are likely to survive on available water. Eat only if there is water to aid chewing and digestion.